THE GLOBAL NAVIGATION SATELLITE SYSTEM

The Global Navigation Satellite System

Navigating into the new millennium

ALESSANDRA A.L. ANDRADE

Routledge

Taylor & Francis Group

LONDON AND NEW YORK

First published 2001 by Ashgate Publishing

Published 2016 by Routledge
2 Park Square, Milton Park, Abingdon, Oxon OX14 4RN
711 Third Avenue, New York, NY 10017, USA

Routledge is an imprint of the Taylor & Francis Group, an informa business

British Library Cataloguing in Publication Data
Andrade, Alessandra A.L.
  The global navigation satellite system : navigating into
  the new millennium. - (Ashgate studies in aviation
  economics and management)
  1.Artificial satellites in navigation 2.Air traffic control
  I.Title
  629.1'3251

Library of Congress Control Number: 2001089778

ISBN 13: 978-0-7546-1825-6 (hbk)

To my parents

Contents

Foreword

Dr. Assad Kotaite, President of the Council,
International Civil Aviation Organization

"The Global Navigation Satellite System" by Alessandra Arrojado Lisbôa de Andrade is a welcome complement to the documentation currently available to the world aviation community on CNS/ATM systems - communications, navigation, surveillance and air traffic management. It strikes a balance between academia and practical realities, as it underscores the pressing need for the world aviation community to renew its commitment to the progressive, systematic and harmonised implementation of CNS/ATM systems world-wide. Indeed, this global project is the cornerstone of an air traffic system for the 21st century which is safe, efficient, environmentally conscious and responsive to the needs of all of the partners in international civil aviation.

The genesis of CNS/ATM goes back to 1972 when ICAO's Seventh Air Navigation Conference recommended that Contracting States pursue the use of satellites for civil aviation. In the early 1980's, the future air navigation system (FANS) concept took shape and, in 1991, the Twelfth Air Navigation Conference approved and recommended the implementation of CNS/ATM systems. Satellite and digital technologies are at the core of CNS/ATM.

Over the past decade, considerable progress was achieved in turning the concept of CNS/ATM systems into reality. ICAO has developed global standards for practically every component associated with the CNS field and is now focussing on the development of operational requirements and procedures for the ATM, through the development of a global ATM operational concept. Substantial guidance material for use by States and by other members of the aviation community has been produced, in response to their specific responsibilities and requirements under the CNS/ATM challenge.

As we move forward, one crucial consideration for the full implementation of CNS/ATM systems will be the creation and adoption of a legal framework for the global navigation satellite systems (GNSS). This is a complex issue and, in order not to delay the implementation of CNS/ATM, the 32nd Session of the Assembly of ICAO adopted, as an interim framework, a Charter on the Rights and Obligations of States relating to GNSS which contains a solemn declaration of the fundamental principles which will govern GNSS in the future. The Assembly also instructed the Council and the Secretary General to consider the elaboration of an appropriate long-term framework to govern the operation of GNSS, including consideration of an international convention for this purpose, and to present proposals for such a framework in time for the 33rd Session of the Assembly in the fall of 2001.

Financing the systems will be another formidable challenge. To assist, ICAO has already put into place guidance material on the establishment and operation, by States or groups of States, of multinational facilities and services which are essential to the implementation of the costly but high capacity components of the CNS/ATM systems. Extensive guidance material has also been developed on related financial issues and cost recovery. Ultimately, there must be convincing business plans for much-needed investments, with special emphasis on the need for clear links between the business cases and operational requirements.

A healthy and growing air transport system contributes to the collective well being of the world's nations; beyond economics, air transport enriches the social and cultural fabric of society and contributes to world peace and prosperity. An efficient, global air navigation infrastructure is essential to the safety and growth of air transport.

A growing understanding of the requirements that dictate the need for the modernisation of such an infrastructure is essential. The careful analysis and meaningful reflection contained in Ms. Andrade's book will facilitate and support the implementation of the new air navigation system of the future, to the benefit of all parties involved in international civil aviation.

Preface

Heralding the beginning of a new era in international civil aviation, the concept of the CNS/ATM systems came into being as the result of joint efforts of the international community, under the aegis of ICAO, in search for a solution to the limitations of the ground-based air navigation systems, which would otherwise inhibit further development of air transport on a global scale. Employing digital and satellite technologies in support of a seamless air traffic management system, the systems are expected to bring improvements upon the present levels of safety, efficiency and accuracy, as well as increased capacity and economic benefits.

As the international community moves forward with the implementation of the systems, their legal implications, institutional framework and financing mechanisms represent a great challenge to States. Particularly, it is their core element, the global navigation satellite system, which promises to be the focus of attention in the new millennium. In a scenario where sovereign States have traditionally been responsible for the provision of air navigation services in their territory, the new satellite-based system suddenly appears to defy the working order, as new practices regarding ownership and control seem to confront the established principle of State sovereignty.

The "Global Navigation Satellite System" not only gives a general overview of the technical aspects of the CNS/ATM systems, but also thoroughly examines the legal, institutional and financing aspects of its navigation component.

This book carefully examines the many concerns raised by States with respect to the availability, continuity and reliability of the air navigation services provided free of charge by the United States' GPS and the Russian

Federation's GLONASS, which systems are under the exclusive control of the individual States. It also focuses on the development of the civilian-controlled Galileo system in Europe, as well as addressing frequency spectrum and orbital position considerations. Particular emphasis throughout is laid on the controversial issue of the allocation of liability in case of damage.

Special consideration is given to issues raised by existing legal tools which bring to bear the compelling need for the development of an appropriate legal framework for GNSS. In considering the fundamental principles to be contained therein, it critically examines the adequacy or viability of an international convention as a long-term means towards providing for legal guarantees which will inspire world-wide confidence in the integrity of the system. It provides detailed examination of all relevant issues, such as liability, certification, administration, financing and cost recovery, as well as future operating structures.

Finally, the book suggests three viable alternatives as recommendations, with different legal and institutional implications, which would eventually help build global confidence in the CNS/ATM systems.

I would like to acknowledge with gratitude all those whose invaluable assistance helped in the completion of this work. I would particularly like to express my very warm thanks to my friend Effie Boikos who diligently read through the manuscript, and especially for her kindness and continued encouragement throughout its preparation. A special word of gratitude goes to Lynn McGuigan at ICAO Headquarters for her unfailing support and unflagging enthusiasm, and for so generously giving me the benefits of her vast knowledge and skills in text processing. Above all my thanks are due to my good friends and fellow colleagues from McGill University, Dr. Ruwantissa Abeyratne and Dr. John Saba, for their constant encouragement and unfailing readiness to offer guidance.

Montreal, June 2001   Alessandra Arrojado Lisbôa de Andrade

List of Abbreviations

AAC	aeronautical administrative communications
ABAS	aircraft-based augmentation system
ACAS	airborne collision avoidance system
ACI	Airports Council International
ADS	automatic dependent surveillance
AEROSAT	Aeronautical Satellite Programme
Allocation	entry in the Table of Frequency Allocations of a given frequency band for the purposes of its use by one or more terrestrial or space radiocommunication services
Allotment	entry of a designated frequency channel in an agreed plan, adopted by a competent conference, through which participating member States distribute among themselves the geostationary orbital positions and radio frequencies
ALLPIRG	all planning and implementation regional groups
AMSS	aeronautical mobile satellite service
AOC	aeronautical operational control
AOR(E)	Atlantic Ocean Region East
APC	aeronautical passenger communications
ARNS	aeronautical radionavigation service
AS	anti-spoofing
ASECNA	Agence pour la Sécurité de la Navigation Aérienne (Africa and Madagascar)
ASM	airspace management
Assignment	an authorisation granted by a national administration for the use of a certain frequency by a radio station

ASTRA Panel	Application of Space Techniques Related to Aviation Panel
ATC	air traffic control
ATFM	air traffic flow management
ATM	air traffic management
ATN	aeronautical telecommunications network
ATS	air traffic services
Band	group of frequencies
C/A code	Coarse Acquisition code
CAA	civil aviation authority
CASITAF	CNS/ATM Systems Implementation Task Force
CNS/ATM	Communication, Navigation, Surveillance/Air Traffic Management
COCESNA	Central American Corporation for Air Navigation Services
DME	Distance Measuring Equipment
DOD	U.S. Department of Defence
ECAC	European Civil Aviation Conference
EGNOS	European Geostationary Navigation Overlay System
ESA	European Space Agency
ESRO	European Space Research Organisation
ETG	European Tripartite Group
EUROCONTROL	European Organisation for the Safety of Air Navigation
FAA	U.S. Federal Aviation Administration
FANS	Future Air Navigation Systems
FANS Phase II	Special Committee for the Monitoring and Co-ordination of Development and Transition Planning for the Future Air Navigation Systems
FARs	FAA's Federal Air Regulations
FIRs	Flight Information Regions
Frequencies	different wave-lengths
FS	fixed services
FTCA	Federal Tort Claims Act
Galileo	European proposal for a constellation of navigation satellites
GASP	Global Aviation Safety Plan
GBAS	ground-based augmentation systems
GEO	geostationary satellite
Global Plan	Global Air Navigation Plan for CNS/ATM Systems
GLONASS	GLObal Navigation Satellite System
GNSS	Global Navigation Satellite System

GPO	DOD's GPS Joint Programme Office
GPS	Global Positioning System
HF	high frequency
IATA	International Air Transport Association
ICAO	International Civil Aviation Organisation
IFATCA	International Federation of Air Traffic Controllers' Associations
IFFAS	International Financial Facility for Aviation Safety
IFR	instrument flight rules
ILS	Instrument Landing System
IMO	International Maritime Organisation
INMARSAT	International Maritime Satellite Organisation
INS	Inertial Navigation System
IOR	Indian Ocean Region
ITU	International Telecommunications Union
JAA	Joint Aviation Authorities
JARs	JAA's Joint Air Regulations
LAAS	Local Area Augmentation System
LACAC	Latin American Civil Aviation Commission
LORAN-C	Long-Range Radio Aids to Navigation
LTEP	Panel of Legal Experts on the Establishment of a Legal Framework with regard to GNSS
MHz	megahertz
MLS	Microwave Landing System
MOU	Memorandum of Understanding
MSAS	Multi-functional Transport Satellite Augmentation System
MSS	Mobile satellite services
MTSAT	Multi-functional Transport Satellite
NAPA	National Academy of Public Administration (U.S.)
NATO	North Atlantic Treaty Organisation
NDB	non-directional beacon
NRC	National Research Centre (U.S.)
OMEGA	OMEGA Navigation System
PANS	Procedures for Air Navigation Services
P-code	Precision code
PIRGS	Planning and Implementation Regional Groups
PPP	public-private partnership
PPS	Precise Positioning Service

Radio waves	eletromagnetic radiation, measured in hertz or cycles, which travels in a straight line at the speed of light and is subject to absorption, diffraction, reflection and diffusion
RAIM	receiver autonomous integrity monitoring
RNAV	area navigation
RNP	required navigation performance
RNSS	radionavigation satellite service
S/A	selective availability
SARPs	ICAO's Standards and Recommended Practices
SBAS	satellite-based augmentations
SDR	special drawing rights
SPS	Standard Positioning Service
SSR	Secondary surveillance radar
TEN	Trans-European Network
TRAINAIR	programme established by ICAO, designed to enhance training effectiveness and efficiency through the use of standardised and modern instructional methodology
UNDP	United Nations Development Programme
USOAP	Universal Safety Oversight Audit Programme
VDL	VHF digital link
VFR	visual flight rules
VHF	very high frequency
VOR	VHF omnidirectional radio range
WAAS	Wide Area Augmentation System
WRC	World Radio Conference

Introduction

Many innovative ideas have been put forward for consideration throughout the years concerning the fundamental features in the development of international air transport policies and the planning of aviation infrastructure. Yet, their particularities and technicalities have long proven not to be guarded against inevitable changes in the aviation environment and the world economy.[1] An assessment of the means whereby they are defined must take into account several factors, economic influences, and particularly, future trends in aircraft movements, passenger and freight traffic flows.[2]

In this regard, there can be no doubt that there is a strong correlation between economic development[3] and air traffic growth.[4] On the one hand, civil aviation plays a catalytic role in the development of the world economy. Besides facilitating local and global communities in terms of leisure and business travel as well as transportation of goods, it also indirectly stimulates economic activities through its end users. A clear example of that might be the undertaking by airline passengers of pre-travel and post-travel purchases of other goods and services, or related expenditures of the freight forwarding business. In addition thereto, airports, commercial airlines and general aviation activities depend on a wide range of inputs from other sources and industries, such as fuel, oil and de/anti-icing fluid suppliers, airframe and engine, avionics and communication equipment manufacturing, maintenance and repair, air navigation services, travel agents, ground-handling services, computer reservation systems, in-flight catering and passenger/cargo facilitation, among others. Suffice it here to say that the direct economic contribution of the civil aviation sector world-wide is estimated at US$ 338 billion, constituted of its value added share, and providing at least 4.1 million jobs, 1.8 million of which within the airlines, about 1.2 million within the aerospace manufacturing industry and at least 1.1 million as direct airport employment.[5]

On the other hand, the pattern of air traffic growth is but a clear reflection of economic conditions experienced over a definite period. International trade developments[6] with the widespread adoption of liberalisation policies have clearly influenced the air transport industry and have had a direct and positive impact on a steadily growing demand for air freight and business travel. The demand for air passenger transportation is primarily determined by income levels and demographics, and the cost of air travel, being partially influenced by the demand for international tourism.[7] Economic cycles, inflation, fluctuations in exchange rates and jet fuel prices affect international travel markets and hence the related demand[8] and subsequent distribution of traffic flows as well as airline yield levels.[9] Finally, traffic growth will vary by geographic region[10] depending on the influence of specific local or regional factors.[11]

As regards aircraft movements, the rapid growth registered in the past decade has not only increased concerns over airport, groundside and airspace congestion, but also continuously affected operations and put pressure on the already hard-pressed airport[12] and air traffic control facilities.[13] As passenger demand increases, air carriers have responded either by scheduling extra flights, by using larger aircrafts, or by managing higher load factors.[14] The air services provided in order to meet such a higher level of demand result from a number of decisions concerning network structure, aircraft type and service frequency, overall largely dependent on the availability of traffic rights, consumer preferences and trade-offs between price and service quality.[15]

Furthermore, the trend towards globalisation and liberalisation in international markets[16] has created important competitive strategies, such as co-operative commercial arrangements among air carriers, alliances, mergers and take-overs, airline consolidation at national levels, low price, frequency and non-stop scheduled-flights, resulting in an increased number of aircraft movements. As a counterpoise, however, it is assumed that the aviation world will witness an increase in the average aircraft size,[17] what might eventually reduce or reverse the pressures to increase frequency at its expense.[18]

The measure of aircraft movements is given in terms of the number of aircraft-kilometres flown or the number of aircraft departures from airports. As such, these measure criteria are extremely relevant for determining the demand for air traffic control facilities, airport planning and other aviation infrastructure. According to recent forecasts, an increase of about 55 per cent and 28 per cent in aircraft-kilometres and aircraft departures respectively is expected between 1995 and 2005.[19] The probabilities are high that it will result in a most serious congestion of airport and airspace alike, an increase in traffic delay and fuel waste and, consequently, in the costs to international civil aviation, compromising the safety of flight. Thus, it is imperious that airport services and infrastructure as well as air traffic control keep pace with the magnitude of the anticipated demand.[20]

In this context, having realised the need to transform their role in the air transport chain, airports, on the one hand, have been reinventing and positioning themselves as centres for economic development, and increasingly, as gateways to growth for their communities and countries. This transformation has expanded beyond their boundaries to encompass surface transport modes, such as high-speed rail connections and road transport terminals, and economic diversification. The demand for the already familiar sites of hotels, industrial parks, business and shopping centres is rising. Likewise, the major socio-economic benefits of airports are being recognised. With a view to catering for the predicted growth and for new aircraft types, the need for planning and providing for airport capacity expansion (slots, gates and terminal capacity) has been duly acknowledged.[21] Moreover, airport capacity and airspace capacity are directly and strongly interconnected, and thus should both be treated as a single integrated system.[22]

On the other hand, as far as airspace capacity is concerned, an analysis of the existing terrestrial-based systems' technology and procedures supporting civil aviation came to expose their shortcomings[23] as regards the capacity to deal with the expected air traffic demand and the future requirements of the civil aviation community. Recognising the challenge represented by the more than ever evident need for global consistency in the provision of air traffic services and for the overcoming of the limitations which would otherwise inhibit further development of air navigation on a global scale,[24] the Council of the International Civil Aviation Organisation[25] established in 1983 the Special Committee on Future Air Navigation Systems (FANS). It was tasked with studying technical, operational, institutional and economic issues, identifying and assessing new concepts and technologies, including satellite technology,[26] and making recommendations for a long-term projection for the co-ordinated evolutionary development of air navigation over a period of the order of twenty-five years.[27]

Having concluded that taking full advantage of existing and foreseen satellite technology would be the key to safe, efficient and orderly evolution in air transport world-wide, a visionary concept which later came to be known as the Communication, Navigation, Surveillance/Air Traffic Management or CNS/ATM Systems was developed.[28] The CNS systems were planned to employ "digital technologies, including satellite systems together with various levels of automation, applied in support of a seamless global air traffic management system".[29] This concept was later endorsed by the Tenth Air Navigation Conference in its Recommendation 9/1,[30] thus signalling the beginning of a new era to international civil aviation and paving the way for its early implementation. The Conference further adopted a series of universally accepted recommendations covering all aspects of CNS/ATM activities, which continue to offer the necessary guidance in the planning and implementation of the systems world-wide.[31]

Indeed, the systems were introduced with a strategic vision, namely "[t]o foster implementation of a seamless, global air traffic management system that will enable aircraft to meet their planned times of departure and arrival and adhere to their preferred flight profiles with minimum constraints and without compromising agreed levels of safety".[32]

Its well-defined mission in coping with the world-wide growth in air traffic demand includes: i) improvements upon the present levels of safety and regularity and upon the over-all efficiency of airspace and airport operations, leading to increased capacity; ii) increase in the availability of preferred flight schedules and profiles; and iii) minimisation of differing equipment requirements.[33]

Notwithstanding the overall benefits[34] expected to be brought about by the CNS/ATM systems, the very nature of their technology is responsible for a major change in the way States will be required to develop and implement air traffic systems in their territories. In a scenario where sovereign States have traditionally been responsible for the procurement, certification, operation and maintenance of their own air navigation systems in accordance with ICAO's Standards and Recommended Practices (SARPs), and air navigation plans, the new satellite-based systems call for a completely new approach on ownership and control in their provision and operation. Particularly, of utmost necessity becomes the interoperability between its elements, which must be ensured so that the goal of a seamless, global navigation and air traffic management can be achieved.[35]

In this regard, the necessity of a smooth transition which should be monitored and co-ordinated by ICAO has been acknowledged so as to guarantee the global planning, harmonisation and implementation of the new systems. Based on the above considerations and following a recommendation[36] of the FANS Committee in its last report, the ICAO Council, in July 1989, established the Special Committee for the Monitoring and Co-ordination of

Development and Transition Planning for the Future Air Navigation Systems (FANS Phase II) with the following terms of reference:

1. To identify and make recommendations for the acceptable institutional arrangements, including funding, ownership and management issues for the global future air navigation system.
2. To develop a global co-ordinated plan, with appropriate guidelines for transition, including the necessary recommendations to ensure the progressive and orderly implementation of the ICAO global, future air navigation system in a timely and cost-beneficial manner.
3. To monitor the nature and direction of research and development programmes, trials and demonstrations in CNS and ATM so as to ensure their co-ordinated integration and harmonisation.[37]

Besides completing its predominant task in regard to the prospective technical architecture of the future CNS system in great richness of detail and flexibility of alternatives, valuable principles for its institutional layout were also developed by the FANS (Phase II) Committee,[38] being widely accepted today that its work, together with that of the previous FANS, will determine the shape of international civil aviation well into the next century.[39]

In continuing to fulfil its mandate under Article 44 of the Convention on International Civil Aviation,[40] ICAO set about to discuss and develop the principles and techniques necessary for international standardisation. Such techniques have mostly been defined, as has significantly progressed the development of material necessary for the planning, implementation and operation of the CNS/ATM systems.[41] SARPs, Procedures for Air Navigation Services (PANS) and guidance material on the various aspects of the systems are largely in place and will continue to be developed in accordance with identified requirements.[42] For example, a new set of GNSS-related SARPs and guidance material has been developed by the Global Navigation Satellite System Panel[43] for inclusion in Annex 10.[44]

The systems, which are both technologically feasible and economically viable,[45] are now in the process of gradual implementation at global, regional and national levels according to ICAO's Global Plan.[46] Regional planning and implementation groups are responsible for the integration and harmonisation of CNS/ATM plans of their various regions, while ICAO will carry out interregional co-ordination to ensure global compatibility of the systems.[47]

The emerging technologies will support a variety of systems designs and implementation options. The challenge for the planner and designer is to develop an adequate understanding of the costs, benefits and operational suitability of these alternatives while considering the legal, organisational and financial aspects; and to orchestrate a co-ordinated programme of ATM improvements that takes into account user needs, their willingness to upgrade their capabilities to achieve operational benefits and also to pay for the changes required by ATM services providers.[48]

The Global Plan, "a living document" which comprises not only technical and operational elements, but also economic, financial, legal and institutional aspects, offers practical guidance to States and regional planning groups on the implementation of the systems, as well as funding and technical co-operation.[49]

At early stages of the discussions, however, many States already expressed their concerns as regards the legal and institutional challenges to be faced in the implementation of the new global CNS/ATM systems. Consideration was especially given to the satellite navigation systems known as Global Navigation Satellite Systems (GNSS),[50] since key components of military roots happen to be currently in control of individual States: the Global Positioning System (GPS) of the U.S. and the GLObal Navigation Satellite System (GLONASS) operated by the Russian Federation. They have both been made available to the international aviation community free of charge for a period of 10 and 15 years, respectively. Although States have generally reacted positively to this initiative, which enabled all, and particularly those without space capabilities, to reap the benefit of satellite-based air navigation facilities,[51] some were filled with apprehension as regards the exercise or loss of sovereign authority,[52] once they would be relying upon signals provided by satellites not under their control. States are particularly concerned with the prospect of GNSS having to be relied upon as the sole means of navigation. Once the traditional ground facilities are dismantled, it has been claimed that the unexpected or unilaterally decided discontinuation of GNSS services might represent the shutting down of the entire air transport system dependant on the use of such services.[53]

Thus, it has been argued that it would be necessary to establish an appropriate global legal framework to govern the operation and availability of GNSS, which would provide from the outset firm guarantees as regards universal accessibility, continuity, accuracy, reliability and integrity, covering also issues of liability and allowing for full participation of all interested parties in the operation and control of GNSS.[54]

Contrary to the views entertained above, others have claimed that the existing legal framework, including the Chicago Convention, is sufficient to govern the system which does not "legally" differ from traditional air navigation aids: "while ... revolutionising global air navigation, they need not revolutionise international aviation law".[55] Arguments in defence of this reasoning state that "a new technological invention does not require legal regulation unless and until it creates specifically new social relations and conflicts of interests" and that "specific legal regulation unavoidably lags behind technological progress, being based on practical experience and needs".[56] Indeed it is difficult to predict which types of legal conflicts may arise, considering that the final configuration of the future systems is yet to be determined.

Still, ICAO has been called upon to consider and develop an appropriate legal framework for GNSS, having the item been given priority in the General Work Programme of the Legal Committee since its 28th Session[57] in May, 1992.

As a first step, taking into consideration both the recommendations of the FANS (Phase II)/3 and of the 28th Session of the Legal Committee, the ICAO Council formulated and adopted in March, 1994, a "Statement of ICAO Policy on CNS/ATM Systems Implementation and Operation",[58] containing general provisions which function rather as policy safeguards than binding principles,[59] but which were nevertheless a strong indication of the incipient global consensus on the desirable general principles for the long-term.[60]

Results of the work carried out by the Panel of Legal Experts on the Establishment of a Legal Framework with regard to GNSS (LTEP),[61] established by the ICAO Council during its 136th Session in 1995, comprised a number of recommendations on the legal aspects related to certification, liability, administration, financing and cost-recovery, as well as future operating structures for GNSS services. It also prepared a Draft Charter on the Rights and Obligations of States Relating to GNSS Services, which embodied certain fundamental legal principals applicable to the implementation and operation of GNSS, including, *inter alia*, the safety of international civil aviation, universal accessibility, continuity, availability, integrity, accuracy and reliability of services, and preservation of State sovereignty.[62]

Those recommendations together with the Draft Charter were presented for information at the World-wide CNS/ATM Systems Implementation Conference held in Rio de Janeiro from 11 to 15 May, 1998. Controversial though it might have been, the Conference supported the adoption of the Draft Charter as an interim measure, while further consideration would be given to the long-term legal framework which, according to the predominant view, should have the form of an international convention.[63] The unique aspect of the Conference in gathering all major partners in civil aviation, from top-level government, industry decision makers and directors of civil aviation authorities to heads of financial institutions and investors, major manufactures, service providers and users was topped out with its addressing of other key issues, such as financial, institutional, technical co-operation and training[64] and the preparation of a "Declaration on Global Air Navigation Systems for the Twenty-first Century".[65]

The work of both the LTEP and the Rio Conference was further considered and endorsed by the 32nd Session of the ICAO Assembly,[66] which also adopted two resolutions,[67] one related to the Charter as an interim measure for the short-term, and a second one regarding the development and elaboration of an appropriate long-term legal framework to govern the implementation of GNSS.

Consideration of the legal framework should not be limited to GNSS only, but also be extended to other aspects of the CNS/ATM systems.[68] In this regard, pursuant to ICAO Assembly Resolution 32-20 and the decision of the Council during its 154th Session on 10 June 1998, a Secretariat Study Group was established to:

a) ensure the expeditious follow-up of the recommendations of the World-wide CNS/ATM Systems Implementation Conference, as well as those formulated by the Panel of Legal and Technical Experts on the Establishment of a Legal Framework with Regard to GNSS (GNSS), especially those concerning institutional issues and questions of liability; and

b) consider the elaboration of an appropriate long-term legal framework to govern the operation of GNSS systems, including consideration of an international convention for this purpose, and to present proposals for such framework in time for their consideration by the next ordinary session of the Assembly.[69]

Further work should not, however, delay implementation of the systems, since there is nothing inherent which is inconsistent with the Chicago Convention.[70]

As Dr. Assad Kotaite, the President of the ICAO Council, has stated:

The trademark of any successful enterprise is its ability to institutionalise the process of monitoring the changing environment, refining its strategy to meet the new imperatives, and modifying its operations accordingly.[71]

In a scenario where the interaction of many different participants is a major concern in the provision, operation, use, financing, management and regulation of the systems, and in which a key navigation component is clearly multifaceted with many different categories of users apart from civil aviation,[72] an important question remains in the air: what are the legal implications of the global navigation satellite system?

Accordingly, in the following pages, a general overview of the CNS/ATM systems is given so as to help in the understanding of their legal and institutional implications. The GNSS will be the focus of this study. Part II is dedicated to the evolution of its existing elements, the emerging elements, as well as frequency spectrum and orbital position considerations. Legal aspects are considered in Part III, where an analysis of the existing legal tools is made and the need and desirability of an international convention is examined. Special attention is given to liability, certification, administration, financing and cost recovery issues, and other fundamental principles in the long-term legal framework for the GNSS. Finally, three viable alternatives, with different legal and institutional implications, are recommended, which ultimately purport to help build global confidence in the ICAO CNS/ATM systems.

Notes

1   For an in-depth analysis of the regulatory responses to the changing air transport world, see ICAO, *Report of the World-wide Air Transport Conference on International Air Transport Regulation: Present and Future*, ICAO Doc.9644 (1994). See generally IATA, *Reinventing the Air Transport Industry - A Vision of the Future*, *Report of the Eight IATA High-Level Aviation Symposium* (1995) [hereinafter *IATA Symposium*]. See especially K. Rattray, "The Changing Regulatory Environment, What Kind of World Will the Airlines be Flying In?" in *IATA Symposium, ibid.*, 22 at 22-31.

2   See ICAO, *Global Air Navigation Plan for CNS/ATM Systems,* version 1 (Montreal: ICAO, 1998), vol. 2 at para.3.1.1.1 [hereinafter *Global Plan*].

3   World gross domestic product (GDP) grew approximately 4.4 per cent in 2000, having amounted to almost 4.0 per cent for the industrialised countries and 5.3 per cent for the developing ones. The Asia/Pacific region, while regaining some of its economic strength with an average of 4.3 GDP growth, was characterised by significant differences between countries, several having suffered from mild to sharp economic recessions the previous years. The Middle East experienced a significant increase of over 6.0 per cent, Africa achieved a 3.6 per cent growth, North America a 5.3 per cent growth and Latin America exhibited a healthier GDP at 4.3 per cent. Recovering from the impact of the recession in the Russian Federation in 1998, Europe experienced the recovery of Eastern and Central regions, achieving a 3.5 GDP growth on average. See ICAO, *Annual Report of the Council – 2000*, ICAO Doc.9770 (2000) at 1 [hereinafter *2000 Council Report*].

4   International forecasts call for a continuation of growth in world-wide air transport: "the total domestic and international air traffic carried by the airlines of the 185 contracting States of ICAO is estimated at about 369 billion tonne-kilometres performed in 1999, an increase of about 6 per cent over 1998. Airlines carried about 1,558 million passengers and some 28 million tonnes of freight in 1999, up from 1,471 million passengers and 26 million tonnes of freight in the preceding year". ICAO

Secretariat, "Annual Review of Civil Aviation – 1999" (2000) 55: 6 *ICAO Journal* 7 at 12 [hereinafter Annual Review].

5   See ICAO, *World-wide CNS/ATM Systems Implementation Conference* (Rio de Janeiro, 11-15 May 1998) [hereinafter WW/IMP], "Impact of Civil Aviation on States' Economies", ICAO WW/IMP-WP/19 (20 March 1998) at 1ff.

6   Reflecting the instability of those economies that are highly exported-oriented and the weak economic performance in various States, world trade volume in goods and services is estimated to have grown at about 4.6 per cent in 1999, compared to almost 10 per cent growth in previous years. See *2000 Council Report, supra* note 3 at 1.

7   According to preliminary results of the World Tourism Organisation, in 2000 some 700 million tourists travelled to foreign countries, having spent about US$475 billion. See *2000 Council Report, ibid.* at 2.

8   Traffic demand may be affected by numerous factors, such as: i) price; ii) frequency; iii) route structure; iv) type of aircraft; v) season; vi) state of the economies of each involved State and vii) the security situation in the destination State. See ICAO, *Manual of the Regulation of International Air Transport*, ICAO Doc.9626 (1996) c. 4.3 at 5 [hereinafter *Manual of Regulation*].

9   See ICAO, *"The World of Civil Aviation, 1999 – 2002"*, ICAO Circ.279 – AT/116 at 7-9. For global trends for airlines and an outlook to the year 2002, see *ibid.*, part 2, at 71-83[hereinafter *World of Civil Aviation*].

10 "On a regional basis, some 36 per cent of the total traffic volume (passengers/freight/mail) was carried by North American airlines. European airlines carried 28 per cent, Asia/Pacific airlines 27 per cent, Latin American and the Caribbean airlines 4 per cent, Middle East airlines 3 per cent and African airlines 2 per cent". *2000 Council Report, supra* note 3 at 2.

11 See *World of Civil Aviation, supra* note 9 at 83. For further information on regional perspectives, trends and forecasts, see *ibid.*, part 3 at 89-117.

12 In 2000, the 25 largest airports in the world (16 of which are located in North America, 6 in Europe and 3 in Asia) handled about 1096 million passengers and approximately 11.5 million commercial air transport movements. This represents about one third of the world total of scheduled and non-scheduled passengers or an average of 114,000 passengers every twenty-four hours as well as an average annual increase of aircraft movements of 2.9 per cent over the 1990-99 period. As regards international air traffic, these airports handled some 532 million passengers, which accounts for about 50 per cent of the world total. See Annual Review, *supra* note 4 at 14; *2000 Council Report, supra* note 3 at 3.

13 See *2000 Council Report, ibid.* at 10.

14 "Load factor is the percentage of available capacity that is actually sold and used by revenue passengers/freight, which can be applied to an aircraft, a route or a sector and expressed for a single sector as, for example, passenger/seats or for multiple sector journeys (taking into account distance) as, for example, passenger/seat/kilometres". *Manual of Regulation, supra* note 8 at 2.

15 See *Global Plan, supra* note 2 at para.3.1.1.5.

16 See generally *IATA Symposium, supra* note 1.

17 For more details on the new large aircraft projects, see *IATA Symposium, ibid.* sess.8 at 125ff.

18 See *Global Plan, supra* note 2, vol. II at para.3.1.1.9.

19   See *ibid.*, vol. II at para.3.1.4.4.
20   See *Global Plan*, *supra* note 2 at para.3.1.4.4.
21   See J.F. Moffatt, "The Airport of the Future", in *IATA Symposium*, *supra* note 1, 102 at 102,103.
22   See *ibid.* at 104.
23   See Chapter 1, below, for more information on the limitations of the current systems.
24   See V.P.Galotti Jr., *The Future Air Navigation System (FANS)* (Aldershot: Ashgate, 1997) at 4 [hereinafter Galotti].
25   Hereinafter referred to as the "Council". Likewise the International Civil Aviation Organisation shall be referred to as "ICAO". ICAO was established as a permanent organisation on 4 April 1947, as an integral part of the Chicago Convention. It became a constituent of the United Nations Organisation and one of its specialised agencies on 1 October 1947. The Assembly, which meets once every three years, is the sovereign body of ICAO; the Council, its governing body, is responsible to the Assembly, and currently composed of thirty-three contracting States. For more information on ICAO, see ICAO, *Memorandum on ICAO, The Story of the International Civil Aviation Organisation,* 15th ed. (Montreal: ICAO, 1994); For details on its aims and objectives, see *infra* note 40.
26   In 1968, following the observations made by the ICAO Communications Divisional Meeting of 1966, the Air Navigation Commission set up a panel of experts to study the applications of space technology relating to aviation (ASTRA Panel), which already identified a potential for global coverage in the system it envisaged. In response to the recommendations of the Seventh Air Navigation Conference, the AEROSAT programme for the launch of an experimental satellite for aeronautical purposes was jointly established by the U.S. Federal Aviation Administration (hereinafter FAA) and the European Space Research Organisation (ESRO), having broken up in 1977 for lack of financial support. An Aviation Review Committee was formed by the AEROSAT Council one year later and directly addressed and recommended ICAO to carry on its work on the characteristics of a future CNS system. See B.D.K Henaku, *The Law on Global Air Navigation by Satellite: A Legal Analysis of the CNS/ATM System* (AST, 1998) at 66-70 [hereinafter Henaku]. See also W. Guldimann and S. Kaiser, *Future Air Navigation Systems: Legal and Institutional Aspects* (Dordrecht: Martinus Nijhoff Publishers, 1993) at 148 [hereinafter Guldimann and Kaiser].
27   See ICAO, Council - 110th Sess., ICAO Doc.9527 – C/1078 C-Min 110 and C-Min 110/9 (1983).
28   See ICAO, *Report of the Fourth Meeting of the Special Committee on Future Air Navigation Systems (FANS),* ICAO Doc.9524 - FANS/4 (2-20 May 1988) [hereinafter FANS/4], Recommendation 2/1 at 2-15. See also Galotti, *supra* note 24 at 4-5.
29   *Global Plan*, *supra* note 2, *Operational Concept and General Planning Principles*, vol. 1. at i.
30   See ICAO, *Report of the Tenth Air Navigation Conference,* ICAO Doc.9583 - AN-CONF/10 (5-20 September 1991), Recommendation 9/1 at 9-3 [hereinafter *AN-CONF/10 Report*].
31   See *Global Plan*, *supra* note 2, vol. 1 at 1.1.9.
32   ICAO, *Global Air Navigation Plan for CNS/ATM Systems Executive Summary* [hereinafter *Executive Summary*].
33   See *ibid.*

34 For an analysis of the expected benefits of the CNS/ATM systems, see Chapter 1, below.

35 See Galotti, *supra* note 24 at 6.

36 See FANS/4, *supra* note 28, Recommendation 5/4 at 5-6.

37 ICAO, *Report of the Fourth Meeting of the Committee for the Monitoring and Co-ordination of Development and Transition Planning for the Future Air Navigation System (FANS PHASE II)*, ICAO Doc.9623 - FANS (II)/4 (15 September – 1 October 1993) at I-1[hereinafter *FANS (II)/4 Report*].

38 See Guldimann and Kaiser, *supra* note 26 at 149.

39 See Galotti, *supra* note 24 at 8.

40 *Convention on International Civil Aviation*, 7 December 1944, ICAO Doc.7300/6; UN Doc.15 U.N.T.S.295, art. 44 (entered into force 4 April 1947) [hereinafter *Chicago Convention*]. Article 44 states that "the aims and objectives of the Organisation are to develop the principles and techniques of international air navigation and to foster the planning and development of international air transport so as to: a) ensure the safe and orderly growth of international civil aviation throughout the world; ... c) encourage the development of airways, airports and air navigation facilities...; h) promote safety of flight in international air navigation..".

41 To take one example, as a follow-up to the work of the FANS Committees, a task force [hereinafter CASITAF] was established to advise the Council on how ICAO could best assist States in the timely and cost-effective implementation of the CNS/ATM systems. See ICAO, *Report of the First Meeting of the Communications, Navigation and Surveillance/Air Traffic Management (CNS/ATM) Systems Implementation Task Force*, CASITAF/1 (24-26 May 1994); ICAO, *Report of the Second Meeting*, CASITAF/2 (20-22 September 1994). Another recent example might be the meetings of the ALLPIRG/Advisory Group which dealt with inter-regional co-ordination and harmonisation mechanism, including the role and scope of PIRGS, facilities and services implementation databases and documents, the year 2K problem, the World Radio Communication Conference, among others. See ICAO, Council, *Report of the Second Meeting of the ALLPIRG/Advisory Group*, PRES AK/594 (11 March 1998). See also ICAO, Air Navigation Services Economics Panel, *Report on Financial and Related Organisational and Managerial Aspects of Global Navigation Satellite System Provision and Operation*, ICAO Doc.9660 (May 1996).

42 See *Global Plan, supra* note 2, vol. 1 at i-1.1.

43 See ICAO, *Third Meeting of the Global Navigation Satellite System Panel*, GNSSP/3 (12-23 April 1999). Significant progress has been made by ICAO in respect to the development of SARPs covering the areas of data link, aeronautical telecommunications network (ATN) and aeronautical mobile satellite services (AMSS). As for the CNS/ATM systems as a whole, the standardisation of equipment requirements and further development of SARPs have been primarily guided by the need to improve safety, efficiency and regularity of flight operations. See ICAO, *World-wide CNS/ATM Systems Implementation Conference Report*, ICAO Doc.9719 (May 1998) at 1.4.2 [hereinafter *WW/IMP Report*].

44 See *Chicago Convention, supra* note 40, Annex 10, Aeronautical Telecommunications, vol. I-V.

45 See A.Kotaite, Opening Address (8th IATA High-Level Aviation Symposium, 24 April 1995) IATA Symposium 4 at 5 [hereinafter Kotaite].

46 The "Global Air Navigation Plan for CNS/ATM Systems" (Global Plan) is an updated and enhanced version of the "Global Co-ordinated Plan of Transition to ICAO CNS/ATM Systems" contained in the FANS (II)/4 Report, *supra* note 37 at 8A-1ff. It comprises two parts: the Operational Concept and General Planning Principles (volume I) and the Global Plan (volume II). Volume I shall provide the necessary guidance for further development of the Basic Operational Requirements and Planning Criteria of the regional air navigation plans (ANPs), as well as the facilities and services required to support implementation of CNS/ATM systems at the regional level. Volume II mainly describes the facilities and services to be provided in order to satisfy the requirements for implementation. See *Global Plan, supra* note 2, vol. 2 at 1.1.

47 See Transition, ICAO CNS/ATM Newsletter 97/3, "ICAO Launches Global Air Navigation Plan for CNS/ATM Systems" (Autumn 1997) at 3.

48 *Executive Summary, supra* note 32 at 5.

49 See *Executive Summary, ibid.*

50 For a detailed description of the entire systems and its components, see Chapter 1, below, at 23.

51 See J. Huang, "ICAO Panel of Experts Examining the Many Legal Issues Pertaining to GNSS" (1997) 52:8 *ICAO Journal* 19 at 19 [hereinafter Huang].

52 See B.D.K. Henaku, "The International Liability of the GNSS Space Segment Provider" (1996) XXIII:I *Annals of Air and Space Law* 145.

53 See J. Huang, "Sharing Benefits of the Global Navigation Satellite System Within the Framework of ICAO" (1996) 3:4 *International Institute of Space Law – Proceedings* 1 at 2.

54 See *WW/IMP Report, supra* note 43 at 5-1-3.

55 *Ibid.* at 5-1-4.

56 M.Milde. "Solutions in Search of a Problem? Legal Aspects of the GNSS" (1997) XXII:II *Annals of Air and Space Law* 195 at 197 [hereinafter Milde].

57 See ICAO, *Report of the 28th Session of the ICAO Legal Committee*, ICAO Doc.9588 – LC/188 (1992).

58 See ICAO, *Statement of ICAO Policy on CNS/ATM Systems Implementation and Operation*, ICAO Doc. LC/29 - WP/3-2 (28 March 1994) [hereinafter *Council Statement*].

59 Huang, *supra* note 51 at 19.

60 See Milde, *supra* note 56 at 200.

61 See ICAO, *Report of the Panel of Experts on the Establishment of a Legal Framework with regard to GNSS*, ICAO Doc.LTEP/1 (23 December 1996) [unpublished][hereinafter *LTEP/1 Report*]; ICAO, *Report of the Panel of Legal and Technical Experts on the Establishment of a Legal Framework with regard to GNSS*, ICAO Doc.LTEP/2 (3 November 1997) [unpublished][hereinafter *LTEP/2 Report*]; ICAO, *Report of the Panel of Legal and Technical Experts on the Establishment of a Legal Framework with regard to GNSS*], ICAO Doc.LTEP/3 (9 March 1998) [unpublished][hereinafter *LTEP/3 Report*].

62 See *WW/IMP Report, supra* note 43 at 5.1.1.

63 See *ibid.* at 5.1.5. See also *Transition, ICAO CNS/ATM Newsletter* 98/05, "Charter or International Convention? Legal Experts Debate" (Autumn 1998) at 2.

64 See R.C.Costa Pereira, Address (42nd Air Traffic Control Association Annual Conference and Exhibits, 30 September 1997) 39:4 *Journal of Air Traffic Control* 56.

65 See *WW/IMP Report*, *supra* note 43, *Declaration on Global Air Navigation Systems for the Twenty-first Century*, at para.7.2 [hereinafter *Rio Declaration*].

66 See ICAO, *Report of the 32nd Session of the ICAO Assembly, Legal Commission*, ICAO, A32/LE (September-October 1998).

67 ICAO, Assembly, 32nd Session, CD-ROM (Montreal, 1998), Charter on the Rights and Obligations of States Relating to GNSS Services, Res. A-32-19 at 64-65 [hereinafter Charter]; *Ibid.*, Development and Elaboration of an Appropriate Long-term Legal Framework to Govern the Implementation of GNSS, Res. A-32-20 at 65-67 [hereinafter Res. A32-20]. See also ICAO Secretariat, "Highlights of the 32nd Assembly" (1998) 53:9 *ICAO Journal* 5 at 9.

68 See *WW/IMP Report*, *supra* note 43 at 5.1.10.

69 ICAO, *Report of the First Meeting of the Secretariat Study Group on Legal Aspects of CNS/ATM Systems*, ICAO SSG-CNS/I-Report (9 April 1999).

70 See *WW/IMP Report*, *supra* note 43, Conclusion 5/1 at 5-2.

71 Kotaite, *supra* note 45 at 4.

72 See P. B. Larsen, "Future GNSS Legal Issues" (Third United Nations Conference on the Peaceful Uses of Outer Space, UNISPACE III, 19-30 July 1999). "[GNSS] provides accurate navigation service for the different modes of transportation, including aviation, water, road, railroad and navigation in outer space. ...GNSS provides positioning for land surveys, agriculture, fisheries, satellite communications, and many other uses, in addition to transportation". But see, J. Huang, Comments on "GNSS - Future Legal Issues", the Discussion Paper presented by P. B. Larsen (UNISPACE III) [unpublished]. "While aviation users may account for a minority of the users of GNSS ... [it] has unique characteristics which differentiate it from other modes of transportation. ... [S]afety of the travelling public is at stake and the risks involved are of a totally different magnitude. Accordingly, consideration of multifunctional GNSS legal principles in the U.N. forum should necessarily take into account the special situation of aviation users and should be closely co-ordinated with the current work of ICAO and perhaps other international organisations such as IMO". For further details on the current applications of GNSS on fields other than aviation, see especially P. Hartl and M. Wlaka, "The European Contribution to a Global Navigation Satellite System" (1996) 12:3 *Space Policy* 167 at 169-170.

PART I

TECHNICAL ASPECTS

1 An Overview of the CNS/ATM Systems

Current Systems — A brief history

It has been said that if aviation pioneers[1] were to return today, "they would readily understand the aerodynamics and propulsion system of a Boeing 747, but they would be completely baffled by the aeroplane's electronic control, navigation and communications equipment".[2]

From a technical viewpoint, this assumption can be easily understood since the basics of aviation technology, at least as far as subsonic flight is concerned, were already well established by the time of the 1944 Chicago International Civil Aviation Conference, and jet engines were already in use in military aircraft.[3]

In the words of the Hon. L. Welch Pogue, former Chairman of the U.S. Civil Aeronautics Board and member of the U.S. Delegation to the Conference, at the close of the Second World War, "as a result of the intense competition for victory" which called for the utmost speed in military travel, significant technological improvements made it possible for aviation to "burst forth from an experimental, crawling promise into an impressive and soaring part of our civilisation".[4]

In those early days, most aircraft were converted military aircraft and powered by piston engines. Although "flying boats" were still relatively common and suitable runways rather few, large four-engine aircraft types, such as the Lockheed Constellation or the Boeing Stratocruiser, dominated long-range flying. Mechanically complex and of questionable reliability, these engines rapidly yielded to the turbine engine, faster and smoother, of which the first to be introduced into commercial service in the 1950s was a turboprop engine, whose overall propulsive efficiency was improved by using its power to drive a propeller. Simpler though they might have been, they were rapidly overtaken by a not much later development, the jet-powered aircraft. At first considered too expensive to operate because of fuel consumption, and extremely noisy, the large, long-range jet aircraft, such as the Havilland Comet, the Boeing 707 and the Douglas DC-8, were soon followed by second-generation types, which entered service during the sixties. Examples of such aircraft are the Boeing 727 and the McDonnell Douglas DC-9. At the next step, there were the commercial supersonic aircraft, a remarkable technical achievement, two types of which were built, the Concorde and the Tupolev Tu-144, as well as the development of the turbofan engine, responsible for an increase in the propulsive efficiency of the jet engine, with a corresponding improvement in fuel consumption. Jumbo jets with larger engines followed, having been designed to cope with much greater passenger capacity, examples being the four-engined Boeing 747 and the three-engined Lockheed L-1011 and DC-10. The latest developments account for very economical, lighter, long-range aircraft with only two engines, such as the Boeing 757 and 767 and also the European Airbus models.[5]

Half a century of major technological progress and the increase in the volume of aviation activity have been accompanied over the years by a substantial development in the vital areas of communications and navigation as well as in its supporting infrastructure (facilities and equipment), namely runways, air traffic control, cockpit instrumentation,[6] control systems, navaids, among others. Despite the continuous improvements, most have been considered inadequate to cope with future demand, as it remains to be seen.

The field of communication in aviation encompasses the operation of navigation aids on the ground, in the air and in space, consisting mainly of radars, landing aids, air to ground and ground to ground telecommunication equipment.[7] Such navigation aids require an extensive use of the radio frequency spectrum, and that is the reason why they are also known as radionavigation aids. Safety of flight requires voice and data communication between the aircraft and air traffic control (ATC), in addition to the interchange of meteorological and flight alerting data, ATC instructions and search and rescue information.[8]

At the beginning, however, air-ground communication had to rely upon radiotelegraphy, since the use of voice communication would not become general practice until after the end of the World War II. Very high frequency (VHF) technology was later employed, but due to its inherent limitations to line of sight distance, high frequency (HF) transmissions, even though not as clear and reliable as VHF, were used in remote areas and over the oceans.

Efforts to improve long-range VHF and HF communication were made over the years, by means of sophisticated antenna systems and single side-band transmissions. Yet, no other major development took place until communications satellites came into existence. Nowadays, wherever there is satellite coverage, voice communication is straightforward for all suitably equipped aircraft. Besides, the systems are able to handle large quantities of digital data for operational purposes.[9]

As regards ground-to-ground communications, connections between ground stations had to be initially undertaken by HF radio band despite its reliability limitations imposed partly by the variability of propagation characteristics. Comprehensive systems of ground links were eventually developed, including under-sea cable and voice communication, and were progressively refined and automated. The replacement of HF voice by satellite communication is considered to be a major step forward in the fixed telecommunications network.[10]

As for navigation, the present systems may be said to encompass three categories: i) very short-range, for approach and landing guidance; ii) short/medium range, for guidance over populated areas, where ground-aids can be provided; and iii) long-range, providing coverage over the oceans or continental areas where ground navigation aids are not available.[11]

The primary approach and landing navigation aid today still is the instrument landing system (ILS), which functions by means of two separate radio beans, capable of defining the approach path in the horizontal and vertical planes, and is associated with three marker beacons, which indicate the distance from the runway. Because of distortions by surrounding areas and interference caused by the relatively narrow frequency band allocated for its use, the need arose for new systems to be designed and a transition plan was established by the ICAO Council in 1987. The implementation of the microwave landing system (MLS), as these other systems are called, has been threatened by the development of the satellite-based navigation systems[12] and its use limited to those locations where it is operationally required and economically beneficial.[13]

As for short/medium-range navigation, the earliest widely used aid was the non-directional beacon (NDB), used for marking points on airways. Other early aids, while providing area coverage, did not confine aircrafts to either fixed or to direct routes. Eventually, the need for channelling aircraft along airways for air traffic control reasons led to the adoption of a second World War point aid, the very high frequency omnidirectional radio range (VOR), as an international standard. But since VOR only provides the pilot with radial information, for the position of the aircraft to be fixed, it is also necessary to provide him with information on its distance from a fixed point, by means of the distance measuring equipment (DME).[14]

Navigational guidance over uninhabited areas and the high seas is generally provided by ground based long-range navigation systems, such as LORAN-C and OMEGA, or by self-contained aids, independent from ground sources, known as inertial navigation systems (INS). While both LORAN-C and OMEGA equipment may be stand-alone, most airborne systems are often duplicated, integrated with other systems and coupled to the autopilot. OMEGA's accuracy depends on the quality of signal reception from the various stations, thus the need for it to be frequently cross-checked with other conventional aids. As for LORAN-C, once highly subject to local interference, it must be limited to areas of good ground wave signal reception. These systems were eventually supplanted by the use of the INS, which is entirely self-contained in the aircraft and operates by sensing the aircraft's accelerations with a gyrostabilised platform. Such information is then integrated by computers to provide accurate position information and navigation data. The system will navigate the aircraft along a predetermined track with waypoints usually inserted prior to departure.[15]

Heavy traffic and low visibility situations gradually led to the need to organise airspace by means of air traffic control, its aim being to promote the safe, orderly, and rapid movement of aircraft through airspace. Since an aircraft may be flown under visual flight rules (VFR) only when visibility is clear enough to allow pilots to visually survey the sky for other traffic, another set of rules was developed to allow flights at any altitude and under all types of weather conditions. However, for aircraft flying under instrument flight rules (IFR), there is a need to follow the controller's directives in order to avoid conflicts with other aircraft.[16]

Because of line-of-sight limitations, the conventional systems controlled aircraft on the basis of their flight plans, which had to be closely adhered to and updated by accurate position reports, so as not to compromise the separation provided. Such ATC situations are known as procedural control.[17] After the introduction of radar coverage, the position of the aircraft would be independently known within the area of coverage, although it was still not possible to identify a particular aircraft. It would only be with the development of the secondary surveillance radar system (SSR) that upon proper interrogation the aircraft receiver would generate a reply signal containing its identification code and altitude.[18] Nevertheless, the application of procedural techniques is still common wherever radar and VHF coverage cannot be provided, particularly in oceanic regions or in low traffic density areas in continental airspace, due to low cost effectiveness. As a consequence, this has required the implementation of carefully controlled track structures over such areas to ensure separation at the expense of optimal flight profiles and system capacity.[19] Once again, the most recent developments in the air control field are based upon the probabilities provided by satellite surveillance and control.[20] Explanation will follow.

Present Shortcomings and Future Benefits

Upon completion of a comprehensive assessment of the characteristics and capabilities of the current air navigation systems and their implementation in various parts of the world, the FANS Committee concluded that the existing air navigation systems suffered from a number of shortcomings in terms of their technical, operational, procedural, economic and implementation nature,[21] which amounted to essentially three factors:

a) the propagation limitations of current line-of-sight systems and/or accuracy and reliability limitations imposed by the variability of propagation characteristics of other systems;
b) the difficulty, caused by a variety of reasons, to implement present CNS systems and operate them in a consistent manner in large parts of the world;
c) the limitations of voice communication and the lack of digital air-ground data interchange systems to support automated systems in the air and on the ground.[22]

Even though the effects of such limitations were not the same for every part of the world and the needs to be satisfied varied considerably due to the different types and densities of traffic, not to mention topography characteristics as well as distinct social and economic conditions,[23] it was recognised that these limitations were inherent to the systems themselves, there being little likelihood that the air traffic service (ATS) system of the time could be substantially improved. New approaches were needed to permit the air traffic management system to be more responsible to the user's needs. Therefore, it was established that the ideal air navigation system would be "a cost effective and efficient system adaptable to all types of operations in as near four dimensional freedom (space and time) as their capability would permit", and which would allow for "considerable improvement in safety, efficiency and flexibility on a global basis".[24] Complementary to certain terrestrial systems, satellite-based CNS systems would be the key to world-wide improvements.[25]

In considering the expected direct benefits from the new CNS systems, communications will see more direct and efficient air-ground linkages, besides improved data handling, reduced channel congestion and communication errors, interoperability across applications and reduced workload.[26]

Improvements in navigation with the full implementation of GNSS include high-integrity, high reliability, all-weather navigation services world-wide and improved four-dimensional navigation accuracy, enabling aircraft to fly in all types of airspace, using simple, on-board avionics to receive and interpret satellite signals. In addition thereto, cost savings from the reduction or non-implementation of ground-based navigation aids are also expected, as well as better airport and runway utilisation.[27]

As regards surveillance, benefits will be derived from reduced error in position reports, surveillance in non-radar airspace, cost savings, improvements in traffic situational awareness, to cite just a few.[28]

An integrated global ATM system will fully exploit the introduction of the CNS technologies, allowing for enhanced safety, increased system capacity, optimised use of airport capacity, reduced delays and diversions, and reduced flight operating costs in terms of fuel consumption and flight crew hours required per flight. Furthermore, it will enable a more efficient use of airspace with more flexibility, reduced separations, dynamic flight planning and accommodation of preferred flight profiles, leading to a reduced controller workload.[29]

In addition thereto, it is expected that the over-all benefits will come to providers and users of the systems alike. To developing States, CNS/ATM particularly provides a timely opportunity to enhance their infrastructure to handle additional traffic with minimal investment. There are also many indirect benefits to be accounted for, such as lower fares and rates, passenger time savings, environmental benefits, increased employment, transfer of high technology skills, industry restructuring and enhanced trade opportunities.[30]

The CNS/ATM Systems

Communications

The communications element in CNS/ATM systems, as envisaged by ICAO, encompasses the complementary use of satellite-based and terrestrial-based technology to provide for global coverage in the exchange of aeronautical data and voice communication between users and/or automated systems.[31]

Capable of carrying both existing categories of aeronautical communications within the allocated frequencies, namely, safety-related and non-safety related communications, still priority shall always be given to safety communication. Requiring high integrity and rapid response, safety-related communication consists basically of ATS communications carried out for ATC, flight information and alerting between ATS units or an ATS unit and an aircraft, and aeronautical operational control (AOC) communications carried out by aircraft operators in relation to safety, regularity and efficiency of flights. Non-safety related communications can be referred to as aeronautical passenger communications (APC), provided by airlines on board the aircraft, and aeronautical administrative communications (AAC) carried out by aeronautical personnel on administrative or private matters.[32]

Routine communications will increasingly take place via digital link rather than the existing channels, therefore reducing the volume of voice communications and, consequently, the work load of pilots and controllers alike. Nevertheless, for non-routine and emergency situations, voice will remain as the primary means of air-ground and ground-ground communication. Transmission of air-ground messages can be carried out over various radio links. However, initially, HF may have to be maintained over polar regions, until suitable satellite coverage is available. The aeronautical mobile satellite service (AMSS), which consists of geostationary satellites[33] offering near global coverage, will be used particularly to provide data link in oceanic and remote continental airspace. Secondary surveillance radar (SSR) mode S data link is specifically suited for surveillance in high-density airspace, while VHF analogue radio will continue to be used in busy terminal areas for voice communication, its efficient use being greatly enhanced with the introduction of different VDL modes.[34]

An aeronautical telecommunications network (ATN) will serve as the infrastructure for global data internetworking, providing for the interchange of digital data between end-users, namely air crew, air traffic controllers and aircraft operators, over dissimilar air-ground and ground-ground sub-networks in support of air traffic services.[35]

Navigation

Designed to provide accurate, reliable and seamless position determination capability, world-wide, by means of satellite-based aeronautical navigation, the navigation element of the CNS/ATM systems has been characterised by the progressive introduction of area navigation (RNAV) capabilities along with the global navigation satellite system (GNSS).[36]

A concept of required navigation performance (RNP)[37] for en-route operations has been approved by ICAO, taking into account a statement of navigation performance accuracy which is expected to be achieved by the population of aircraft within a given airspace. It has been extended to cover approach, landing and departure operations, having been defined in terms of required accuracy,[38] integrity,[39] continuity[40] and availability[41] of navigation. RNAV operations within the RNP concept facilitate a flexible and more direct route structure, circumventing the need to fly directly over terrestrial-based navigation facilities, and allowing equipped aircraft to adhere closely to their preferred flight paths within prescribed accuracy tolerances.[42] In general terms, RNAV equipment operates by automatically determining aircraft position using input from different sources, such as VOR, DME, INS systems, satellites being amongst the most recent ones.

It should be noted that whereas it used to be common practice to prescribe the mandatory carriage of certain equipment to indicate required navigation performance capability, there are no restrictions whatsoever as to how RNP requirements are to be met, so that compliance can be achieved by the provider State or the aircraft operator through the use of any suitable navigation system.[43] However, operational approval in the various RNP-type airspaces by the State of the operator is necessary and should be granted for each individual operator as well as for each individual aircraft type used. Approval procedures have been developed by a few States only and for specific applications. Therefore, the need arises to ensure co-ordination and absolute compatibility between States in the definition of certification and approval requirements for users, infrastructure responsibilities for providers and training requirements, so that international standards can be achieved.[44]

Amidst independent navigation systems which potentially could meet the requirements for sole means navigation, a fundamental element is revolutionising air navigation: the Global Navigation Satellite System.[45] GNSS can be defined as a global position and time determination system comprising one or more satellite constellations, a ground segment, aircraft receivers, and system integrity monitoring, augmented as necessary to support the RNP for the actual phase of navigation.[46] Based on satellite ranging, position is determined by processing range measurements to at least four satellites in view used as reference points. Radio signals being transmitted provide each satellite's position and the time of the transmission. The system works by timing how long it takes a signal to reach the GNSS receiver and then calculating a distance from that time. Thus, it may be used to determine the three-dimensional geographic co-ordinates and real-time position of an aircraft, the course and distance to the destination, and deviation from the desired track.[47]

As mentioned earlier in this work, two systems are presently in operation, namely, the Global Positioning System (GPS) of the U.S. and the Global Navigation Satellite System (GLONASS) of the Russian Federation. Both systems were originally designed and operated as military positioning systems, having been offered to the international community as a means to support the evolutionary development of the GNSS.[48] Through an exchange of letters,[49] in 1994 the ICAO Council accepted the U.S.' offer for a minimum period of 10 years free of charge. In 1996, it also accepted a similar offer by the Russian Federation for a period of 15 years.

The current GPS constellation consists of twenty-nine operational satellites (Block II, IIA, IIR) in six orbital planes, operating in near-circular 20 200 km orbits at an inclination angle of 55 degrees to the equator, each one completing an orbit in approximately 12 hours. As for the ground-based control segment, five Monitor Stations, located at Ascension Island, Colorado Springs, Diego Garcia, Hawaii and Kwajalein, track the satellites in view and accumulate ranging data, which is then processed at a Master Control Station in Schriever Air Force Base, Colorado. Updated information on each satellite's navigation messages (satellite clock and orbit states, system time and status messages) is transmitted to the satellites via ground antennas.[50]

The GLONASS space segment is composed of twenty-four satellites in three orbital planes, inclined 64.8 degrees, at an altitude of 19 100km, with an orbital period of approximately 11 hours and 15 minutes. In a similar manner to GPS, the control segment works by means of a System Control Centre located in the Moscow region and a number of Command Tracking Stations spread throughout the country, which are responsible for tracking GLONASS satellites in view and accumulating ranging data and telemetry from their signals.[51]

The user segments of both navigation systems consist of antennas and receivers which provide positioning, velocity and precise time to users.

Despite offering several advantages over currently available terrestrial-based navigational systems, there are many relevant issues of concern to the international community as regards the use of GPS and GLONASS for navigation purposes, particularly because of important public safety considerations. Firstly, due to inherent limitations, neither system is capable on its own of meeting the RNP requirements for all phases of flight. Besides having limited ability to warn users of malfunctions, what might reveal integrity problems, the accuracy levels afforded are lower than required for the more stringent phases of flight, particularly those associated with precision approaches and landing operations. Potential continuity and availability obstacles must be given due consideration as well. Lastly, legitimate concerns faced by States also reflect the institutional commitment of the signal-providers to maintain reliable services available to the international community, as stated in the instruments exchanged and related policy declarations, as well as the lack of international control.[52]

As a result, various degrees of GNSS augmentation are required to ensure the complete safety of operations, namely aircraft-based (ABAS), ground-based (GBAS) and satellite-based augmentations (SBAS).

There are different types of aircraft-based augmentation techniques, the most important of which being the so-called receiver autonomous integrity monitoring (RAIM), whereby an airborne GNSS receiver autonomously monitors the integrity of the navigation signals from GNSS satellites. Multiple independent positions may be computed and compared, so that a faulty satellite giving incorrect information can be detected and excluded, once such positions do not match. Other reliable techniques employed whenever there are insufficient satellites with suitable geometry in view include inertial systems and altimetry aiding.[53]

GBAS provides differential information, locally or within a small region, by means of a monitor located at or near the airport where precision operations are desired. Signals providing corrections to enhance position accuracy as well as integrity information are transmitted directly to aircraft in the vicinity via a line-of-sight data link.[54]

There can be no doubt, however, that the most practical means to provide augmentation coverage over large areas is through the use of satellites. Its simplest form is the broadcast of satellite integrity status via a geostationary satellite. It has been said, however, that the provision of SBAS by geostationary satellites has certain limitations and therefore cannot be expected to support all phases of flight, especially precision approach and landing of higher categories. For differential coverage over an extensive geographical area, the wide area augmentation is used. It involves networks of data collection ground stations, usually separated by more than 1000km, and located at precisely surveyed locations, where information is collected, and then transmitted to a central facility, there to be processed to derive corrections related to satellite clock, ephemeris and ionospheric delay. Such information is subsequently uplinked to a geostationary satellite constellation and then broadcast to users within the coverage area on the same frequency as the navigation system concerned.[55]

Surveillance

As explained above, from a cost effective standpoint, and in the absence of a single system capable of meeting all defined surveillance performance requirements, several surveillance systems with different characteristics and capabilities are presently necessary to handle the extremely varied traffic conditions, and may be flexibly used as a stand-alone or in combination, provided they meet the set parameters for an operational scenario in a given airspace.[56] Under proper terminology, these systems are known as dependent surveillance and independent surveillance systems.

Representing the first category mentioned, the voice position reporting consists in a system whereby the position of an aircraft is determined from on-board equipment and then conveyed by the pilot to ATC by VHF and/or HF radios. It is expected it will continue to be used in oceanic airspace as well as in area control service outside radar coverage.[57]

Independent surveillance, on the other hand, is based on radar. Although traditional SSR, whose functioning has already been described in this chapter, will continue to be used in the CNS/ATM environment, it is the use of its Mode S that deserves special attention. This enhanced technique not only permits the selective interrogation of suitably equipped aircraft, therefore eliminating garbling, but also two-way data links between Mode S ground stations and transponders, hence constituting the appropriate surveillance tool for terminal areas and high-density airspace. On the other hand, the use of primary radar is already rapidly declining, although it will still continue for a variety of national applications, including weather detection.[58]

The major breakthrough, however, has been the introduction of automatic dependent surveillance (ADS) for use in areas where radar-based surveillance is not feasible or as an adjunct or back-up for such systems. Using ADS, aircraft will automatically transmit their position and data derived from on-board navigation, via satellite or other digital communication links, to an ATC unit. Software is currently being developed to enable the direct use of this data by ground computers to detect and resolve conflicts. As an expansion of the ADS technique, another concept has been developed, the so-called ADS-broadcast, by means of which aircraft will periodically broadcast their position to other aircraft as well as to ground systems.[59]

Lastly, providing advice to the pilot on potential conflicting aircraft is the airborne collision avoidance system (ACAS), based on SSR transponder signals which operate independently from the ground. An enhanced version, ACAC III, expected to generate both horizontal and vertical resolution advisories, is currently under development.[60]

Air Traffic Management

It has been stated that "the primary goal of an integrated ATM system is to enable aircraft operators to meet their planned times of departure and arrival and adhere to their preferred flight profiles with minimum constraints and no compromise to safety"[61] in the most optimum and cost-efficient manner.

Maximum flexibility with guaranteed safe separation, this is the basic precept behind the concept known as "free flight".[62] Essentially envisaging the establishment of a single and continual airspace characterised by the absence of designated air route networks, it allows aircraft to make full use of all available airspace by flying the most efficient route between two points, being it either "the shortest route to minimise flight time, or a route that takes advantage of favourable wind and weather patterns to minimise fuel burn".[63]

Free access to airspace cannot be taken for granted. The Chicago Convention has clearly stated that no scheduled international air service may be operated over or into the territory of a contracting State, except with the special permission or other authorisation of that State, and in accordance with the terms of such permission or authorisation.[64] Moreover, airspace is a shared resource between civil and military users. Each contracting State may, for reasons of military necessity or public safety, restrict or prohibit uniformly the aircraft of other States from flying over certain areas of its territory.[65] Consequently, information on planned movements of civil aircraft and their intended flight path in real time has to be made available to military units.[66]

Therefore aircraft operators would still be required to file flight plans and designate their chosen routes, so that controllers might access whether there are likely to be any conflicts en-route, which means that the traditional functions of ATC must continue to be provided as part of a global ATM system.[67]

Whereas the overall effectiveness and feasibility of free flight[68] remains to be proven, the advancements in CNS technologies will serve to support ATM accomplish this goal. Nevertheless, in order to take full advantage of the new capabilities, an evolutionary transition process is required. Improvements must keep pace with user needs and will favour implementation in contiguous regions. International harmonisation of ATM standards and procedures is essential for integration into a regional and global ATM network.[69]

The envisaged ATM system encompasses several elements, all of which must be fully interoperable and integrated into a seamless, global system, where airborne and ground capabilities are linked and used together. These elements are airspace management (ASM), air traffic services (ATS), air traffic flow management (ATFM) and ATM-related aspects of flight operations.[70]

The concept of ASM reflects not only the sharing of airspace between military and civil users, but also the flexibility of airspace to accommodate ATM requirements for CNS operations. It also includes infrastructure planning as regards airspace organisation, services and facilities, as well as separation minima, with the objective of facilitating the optimal use of airspace, with increased safety and efficiency.[71]

The primary element of ATM will continue to be ATS. Such services are provided by ground facilities, usually operated by national civil aviation authorities or international air traffic organisations, such as the European Organisation for the Safety of Air Navigation (EUROCONTROL). Nevertheless, there has been a growing tendency to have operational services transferred to autonomous authorities.[72]

ATS itself is composed of three elements: i) the flight information service, responsible for the provision of useful information for the safe and efficient conduct of flights; ii) alerting services, which serve to notify appropriate bodies regarding aircraft in need of search and rescue operations; and iii) air traffic control, whose functions include the prevention of collisions between aircraft, and obstructions in the manoeuvring area, while expediting and maintaining an orderly traffic flow.[73]

ATFM is a necessary complement to ATS, which basically aims to optimise air traffic flows, reducing aircraft delay so as to help prevent system overload which might otherwise carry serious safety implications.[74]

Lastly, suffice it here to say that for CNS systems to provide maximum benefits through enhanced ATM, access to global meteorological information on a far shorter time scale than has been customary is required. So is essential the support of aeronautical information services. Many States have begun developing electronic aeronautical databases to improve the speed, efficiency and cost-effectiveness of aeronautical information.[75]

Human Factors and Training Needs

It has been acknowledged that "CNS/ATM systems are technology-intensive, and their safest and most efficient performance is predicated upon the correct utilisation of technology, as intended by its designers".[76] Nevertheless, the much higher levels of automation introduced with new technology along with the interdependency of the systems' elements have raised additional and most serious challenges in respect to human factor issues.[77]

In principle, automation should allow for increased efficiency and safety of operations and help prevent errors[78] by diminishing direct and active human involvement in systems operation. However, the role of technology in the actual fostering of human error has often been absolutely overlooked.

> Most accidents involving high technology systems appear to be mis-operations of technological systems that are otherwise fully functional, and are therefore labelled as human error. The typical belief is that the human element is separate from technology, and that problems reside therefore either in the human or in the technical part of the system. The view ignores, among other things, the role of human capabilities and limitations, and the pressures that the system's production objectives impose upon operational personnel.[79]

Whereas close interaction with technology is necessary, most of the subsequent problems are essentially related to deficient human-machine interface. The most important human factors issue in this regard is the ability of the human operator to maintain situational awareness. For example, in "mode error" situations, there happens a joint human-machine breakdown, in which the human element unintentionally loses track of the machine configuration and the machine simultaneously misinterprets his inputs.[80]

There are but two alternatives to address such problems with different financial implications: i) during the design stage of the system; or ii) after its implementation in the operational context. Traditionally, remedial actions taken after the identification of shortcomings on human performance have been the preferred path, though incurring continuous expenses for training on a routine basis. In order to maximise safety and cost-effectiveness of CNS/ATM systems, a proactive management of human factors is therefore advised, even if it might initially incur additional costs, for those will be paid only once in the system's lifetime.[81]

The standing policy of ICAO on human factors has been established in the Assembly Resolution A32-14 (Appendix W), which resolves that:

1. Contracting States should take into account relevant human factor aspects when designing or certifying equipment and operating procedures and when training and/or licensing personnel;
2. Contracting States should be encouraged to engage in far-reaching co-operation and mutual exchange of information on problems related to the influence of human factors on the safety of civil aviation operations; ...[82]

As a direct consequence of all of the above-mentioned, training is to play a fundamental role in CNS/ATM systems implementation as a significant but essential investment. Moreover, a seamless global navigation system will require a prepared international team and a consistent, quality level of training throughout the world.[83] Major changes to civil aviation job profiles are expected to happen, so foundation training in the basic concepts and technologies is required along with co-ordination of training development at the regional level. A programme designed to enhance training effectiveness and efficiency through the use of standardised and modern instructional methodology has been established by ICAO under the name of TRAINAIR with the strong support of the United Nations Development Programme and should be made the widest use of. It prepares high quality standardised training material for sharing between training centres and assists in the global co-ordination and harmonisation of training development.[84]

Human factors SARPs have been developed by ICAO addressing issues relevant to the certification process of equipment, procedures and personnel.[85]

Finally, the assertion that safety of international civil cannot – ever – be taken for granted is of paramount importance and should be given overriding priority. Although early implementation is being pushed forward by airlines and provider States, who legitimately expect to see returns for the investments in airborne equipment and systems they have made, an incremental implementation program with regular evaluations and validations is absolutely necessary to guarantee safety, allowing time for proper ATC training and licensing process.[86]

The view expressed above has obvious legal implications, especially as regards the proper allocation of liabilities in the event of an aircraft accident. The question may arise whether a reallocation of responsibilities between the pilot in command and air traffic control, as well as the aircraft operator will be a necessary consequence so as to legally reflect the changed interface between all participants in the CNS/ATM systems.[87]

Notes

1   Although the Wright brothers are world-renowned for the first alleged human flight, it was Alberto Santos Dumont, of Brazilian nationality, who achieved the world's first publicly performed, and properly verified, recorded and monitored mechanical flight under technical conditions, using a "heavier-than-air" machine which was built by himself and named "14-Bis". On 23 October 1906, at the Bagatelle field, in Paris, he flew a distance of 60 metres at a height varying between 2-3 metres. On 12 November, flying against the wind, he made his famous 220-metre flight at an altitude of 6 metres in 22 1/5 seconds, for which feat he was duly awarded the Aéro Club de France Prize for the first aircraft that, taking off under its own power should cover a distance of 100 metres with no more than 10 per cent variation from level flight. See Ministério da Aeronáutica, *Alberto Santos Dumont, The Father of Aviation,* (Brazil: Editorial Antártica, 1996) at 26-29. See also A. J. Marchand, "Santos-Dumont: Pionnier de l'Aviation" (1996) 77:4 *AeroFrance* 4-6.

2   L. Mortimer, "1944 – 1994, A Half Century of Technological Change and Progress" (1994) 49:7 *ICAO Journal* 33 at 33 [hereinafter Mortimer].

3   See *ibid.*

4   L.Welch Pogue, "The International Civil Aviation Conference (1944) and Its Sequel: The Anglo-American Bermuda Air Transport Agreement (1946) – Appendix 1, "The Manifest Destiny of International Air Transport" (1994) XIX:I *Annals of Air and Space Law* 3 at 3, 43-44.

5   See Mortimer, *supra* note 2 at 33-36.

6   As far as automated flight control is concerned, a recent development is the Flight Management System. When loaded with a complete flight plan, in conjunction with the autopilot, it is able to conduct all phases of flight, from take-off to landing, without human pilot's intervention. See Mortimer, *supra* note 2 at 38.

7   See V.P.Galotti Jr., *The Future Air Navigation System (FANS)* (Aldershot: Ashgate, 1997) at 53 [hereinafter Galotti].

8   See B.D.K Henaku, *The Law on Global Air Navigation by Satellite: A Legal Analysis of the CNS/ATM System* (AST, 1998)at xv [hereinafter Henaku].

9   See Mortimer, *supra* note 2 at 41.

10   See Galotti, *supra* note 7at 55.

11   See Mortimer, *supra* note 2 at 42.

12   See *ibid.*

13   See Galotti, *supra* note 7 at 103.

14   See Mortimer, *supra* note 2 at 42.

15   See Galotti, *supra* note 7 at 99.

16   See S.K. Hamalian, "Liability of the U.S. Government in Cases of Air Traffic Controller Negligence" (1996) XI *Annals of Air and Space Law* 58.

17   See Galotti, *supra* note 7 at 143.

18   See Mortimer, *supra* note 2 at 44.

19   See Galotti, *supra* note 7 at 143.

20   See Mortimer, *supra* note 2 at 44.

21   See ICAO, *Global Air Navigation Plan for CNS/ATM Systems,* version 1 (Montreal: ICAO, 1998), vol. 1 at para.1.2.1. [hereinafter *Global Plan*].

22   ICAO, *Report of the Fourth Meeting of the Special Committee on Future Air Navigation Systems (FANS)*, ICAO Doc.9524 - FANS/4 (2-20 May 1988) at para.2.1.1 [hereinafter FANS/4]; ICAO, *Report of the Tenth Air Navigation Conference*, ICAO Doc.9583 - AN-CONF/10 (5-20 September 1991), at 2A-1, para.2.1[hereinafter *AN-CONF/10 Report*].

23   See FANS/4, *ibid.*

24   *Global Plan, supra* note 21, vol. 1 at para.1.2.2.

25   See AN-CONF/10 Report, *supra* note 22 at para.3.1.

26   See *Global Plan, supra* note 21, vol.1 at paras.1.3.2.2 and 1-7.

27   See *ibid.* See also A. Delrieu, "CNS/ATM: le Concept et le Système tel qu'Adoptés para L'OACI" (1995) 13 *Le Transpondeur* 4 at 7.

28   See Global Plan, *supra* note 2, intro. at para.1-7.

29   See *ibid.* at para.1.3.5 and 1.7; L. Turner, "Transitioning to CNS/ATM – Tools to the Future" (1997) 39:3 *Journal of Air Traffic Control* 13 at 15.

30   See *Global Plan, supra* note 21, vol. 1 at paras.1.4.3, 1.4.6 and 1.4.9.

31   See *Global Plan, ibid.* at para.5.1.1.

32   See *ibid.* See also Henaku, *supra* note 8 at 74-75.

33   A geostationary satellite (GEO) is a geosynchronous satellite whose circular and direct orbit lies in the plane of the earth's equator. This orbit, the so-called geostationary orbit, is located at an altitude of approximately 35,786.557km above the earth's surface. With a period of revolution equivalent to the rotation of the earth, a satellite there placed appears stationary in relation to a point on earth. Station keeping operations, however, are necessary to keep it at the desired position, since it may be affected by various natural forces. See R. Jakhu, "The Legal Status of the Geostationary Orbit" (1982) 7 *Annals of Air and Space Law* 333 at 333, note 1. For more information, see J.Wilson, "The International Telecommunication Union and the Geostationary Orbit - An Overview" (1998) XXIII *Annals of Air and Space Law* 241.

34   See *Global Plan, supra* note 21, vol.1 at paras.5.3, 5.4 and 5.5; *AN/CONF 10 Report*, *supra* note 22 at para.3.2.1.

35   See *Global Plan, ibid.* at para.5.6.1; ICAO, *Global Air Navigation Plan for CNS/ATM Systems Executive Summary* at 5 [hereinafter *Executive Summary*].

36   See *Global Plan, ibid.* at para.6.1.1.

37   Navigation system performance requirements have been defined by ICAO in the *Manual on Required Navigation Performance* (ICAO Doc.9613) and the *RNP Manual for Approach, Landing and Departure* for a single aircraft and for the total system, including the signal-in-space, the airborne equipment and the ability of the aircraft to fly the desired trajectory. See ICAO, *Report on the Third Meeting of the Global Navigation Satellite System Panel*, Appendix C to the Report on Agenda Item 1, GNSS/3-WP/66 (12-23 April 1999) at para.C.3.1.1 [unpublished][hereinafter *GNSSP Report*]. Consideration must be given to a number of factors in order to determine the appropriate requirements for a particular region, e.g. the traffic density, the complexity of the airspace, and the existing possibility of air traffic control intervention. See *GNSSP Report, ibid., Report on Agenda Item 1* at para.1.2.3.

38   "GNSS position error is the difference between the estimated position and the actual position. For any estimated position at a specific location, the probability that the position error is within the accuracy requirement should be at least 95 per cent". *Ibid.* at para.C.3.2.1.

39   "Integrity is a measure of the trust which can be placed in the correctness of the information supplied by the total system. [It] includes the ability of a system to provide timely and valid warnings to the user (alerts) when the system must not be used for the intended operation (or phase of flight)". *Ibid.* at para.C.3.3.1.

40   "Continuity of a system is the capability of the system to perform its function without non-scheduled interruption during the intended operation". *Ibid.* at para.C.3.4.1.

41   "The availability of GNSS is the portion of time during which the system is to be used for navigation during which reliable navigation information is presented to the crew, autopilot, or other system managing the flight of the aircraft". *Ibid.* at para.C.3.5.1.

42   See especially Galotti, *supra* note 7 at 111-119. See also Henaku, *supra* note 8 at 170-171; *Global Plan, ibid.*, vol. 1 at 6.2; M.C.F.Heijl, "CNS/ATM Road Map for the Future" (1994) 49:4 *ICAO Journal* 10 at 10 [hereinafter Heijl].

43   See Galotti, *ibid.* at 111-113.

44   See D. Moores, "RNP Implementation Demands Commitment and Careful Consideration of Many Issues" (1998) 53:2 *ICAO Journal* 7 at 8-9.

45   For detailed technical information, see ICAO, *Guidelines for the Introduction and Operational Use of the Global Navigation Satellite System*, ICAO Circ.267.

46   See *Global Plan, supra* note 21, vol. 1 at 6.3.1.

47   See Galotti, *supra* note 7 at 105; J. Huang, "ICAO Panel of Experts Examining the Many Legal Issues Pertaining to GNSS" (1997) 52:8 *ICAO Journal* 19 at 19.

48   See WW/IMP, "GNSS System Status and Standardisation in Progress", ICAO WW/IMP-WP/36 (11 May 1998) at para.2.2.

49   See Letter from D. Hinson, FAA Administrator, to A. Kotaite, President of ICAO Council (14 October 1994); Letter from A. Kotaite to D. Hinson (27 October 1994), ICAO State Letter LE 4/4.9.1-94/89, attachment 1 (11 December 1994); Letter from N.P. Tsakh, Minister of Transport of the Russian Federation, to A. Kotaite, President of ICAO Council (4 June 1996), Letter from A. Kotaite to N.P. Tsakh (29 June 1996), ICAO State Letter LE 4/49.1-96/80 (20 September 1996) [hereinafter Letters].

50   See U.S., *Global Positioning System Data and Information Files* (U.S. Naval Observatory, Automated Data Service), http://tycho.usno.navy.mil/gps.html (date accessed 12 August 2000) [hereinafter USNO Data Service]. For additional information concerning GPS, see U.S., *Global Positioning System Standard Positioning Service – Signal Specification*, 2ⁿᵈ ed. (The U.S. Coast Guard, 1995). See also Chapter 2, below.

51   See Russian Federation, Ministry of Defence, *Global Navigation Satellite System GLONASS* (Moscow: Scientific Co-ordination Information Centre, 2000), http://mx.iki.rssi.ru/SFCSIC/english.html (date accessed: 12 August 2000); WW/IMP, "Results of GNSS Assessment For Application in Approach, Landing and Departure", ICAO WW/IMP-WP-37 (11 May 1998) [hereinafter WW/IMP-WP-37], Appendix at para.1.2.1. For additional information on GLONASS, see Russian Federation, Ministry of Defence, *GLONASS Interface Control Document*, version 4.0 (Moscow: Scientific Co-ordination Information Centre, 1998). See also Chapter 2, below.

52   See Galotti, *supra* note 7 at 107. See also N. Warinsko, "Du GPS au GNSS, Le Point sur la Situation Internationale" (1995) 13 *Le Transpondeur* 19 at 20-21.

53   See *Global Plan, supra* note 21 at para.6.4.2; WW/IMP-WP-37, *supra* note 51 at 1.3 ff.

54   See *Global Plan, ibid.* at para.6.4.3.1.

55   See WW/IMP-WP-37, *supra* note 51 at 1.2.7 and 1.2.10.

56   See *supra* note 18 and accompanying text.

57   See *Global Plan, supra* note 21, vol. 1 at para.7.2.1.

58   See, *ibid.* at paras.7.2.3.1, 7.2.2.1; *AN-CONF/10 Report, supra* note 25 at 3.2.3.

59   See *Executive Summary, supra* note 35 at 6; WW/IMP, "Surveillance Systems", WW/IMP-WP/40 (11 May 1998).

60   See *Global Plan, supra* note 21, vol. 1 at para.7.5; WW/IMP, *ibid.*, "Airborne Collision and Avoidance Systems", ICAO WW/IMP-WP/41 (11 May 1998).

61   *Global Plan, ibid.* at para.4.2.2.1.

62   Free flight is defined as the operating capability under instrument flight rules (IFR) in which aircraft operators may select their preferred flight paths and speed in real time. Any air traffic restriction will only be imposed as long as it is necessary to e.g. a) ensure separation; b) preclude exceeding airport capacity; c) prevent unauthorised flight through special airspace; or d) ensure safety of flight. In any event, such restrictions are to be limited in extent and duration to correct the identified problem. See A.Paylor, "Free Flight – The Ultimate Goal of CNS/ATM?" in ISC/ICAO, Integrating Global Air Traffic Management (London: ISC, 1997) 120 at 122.

63   See *ibid.*

64   *Convention on International Civil Aviation*, 7 December 1944, ICAO Doc.7300/6; UN Doc.15 U.N.T.S.295, art. 6 (entered into force 4 April 1947) [hereinafter *Chicago Convention*].

65   *Ibid.* art. 9.

66   See M.C.F.Heijl, "Aviation Community Working on the Development of Infrastructure Needed to Support Free Flight" (1997) 52:3 *ICAO Journal* 7 at 8.

67   *Ibid.*

68   Potential factors which could hamper realisation of the full benefits of free flight include airspace congestion at centralised crossing points and limited airport (runway and terminal area) capacity.

69   See *Global Plan, supra* note 21, vol. 1 at para.4.2.2.3ff.

70   See *ibid.* at para.4.3.8.

71   For a practical example of the application of airspace planning methodology, see Heijl, *supra* note 42 at 11.

72   See *Global Plan, supra* note 21, vol. 1 at 4.1.1.

73   See *Global Plan, ibid.,* vol. 1 at 4.3.8.18.

74   See Heijl, *supra* note 42 at 12.

75   See *Executive Summary, supra* note 35 at 7. For detailed information, see *Global Plan, supra* note 21, vol. 1, ch. 8 and 9.

76   ICAO, *World-wide CNS/ATM Systems Implementation Conference Report*, ICAO Doc.9719 (May 1998) at 6.2.1 [hereinafter *WW/IMP Report*]; ICAO Secretariat, "Increased ATC Automation May be Inevitable to Handle Increasing Traffic and Data" (1993) 48:5 *ICAO Journal* 16 at 16-17.

77   See N.Vidler, "Human Factors Aspects in CNS/ATM Systems" (1996) 38:3 *Journal of Air Traffic Control* 72 at 73 [hereinafter Vidler].

78   Human participation in automated systems operations comprises not only the monitoring of the systems but also its manual take-over in case unexpected operational conditions, which might come to jeopardise safety, are observed. Nevertheless, there might happen that the human element is not trained for that particular situation which, on the other hand, was also not anticipated in the design phase of the system. See

WW/IMP, "Human Factors Issues in CNS/ATM", ICAO WW/IMP-WP/30 (11 May 1998) at para.2.4.

79 WW/IMP, *ibid.* at para.2.5.

80 See *Transition, ICAO CNS/ATM Newsletter* 98/05, "Human Factors and Training: Crucial Issues in CNS/ATM Implementation" (Autumn 1998) at 1.

81 See WW/IMP, "ICAO Global Strategy for Training and Human Factors", ICAO WW/IMP-WP/13 (11 May 1998) [hereinafter WW/IMP-WP/13] at paras.3.1.4 ff; WP/30 at paras.3.7ff and 4.2; *WW/IMP Report, supra* note 76, Conclusion 6/2.

82 ICAO, *Assembly, 32nd Session, Consolidated Statement of ICAO Continuing Policies and Associated Practices Related Specifically to Air Navigation, Appendix W, Flight safety and human factors*, Res. A32-14.

83 See *WW/IMP Report, supra* note 76, Conclusion 6/5.

84 See M.A.Fox, "ICAO Ready to Help Meet Global Training Needs Associated with the CNS/ATM Systems" (1995) 50:4 *ICAO Journal* 14 at 14ff; *Global Plan, supra* note 21, vol. 1 at para. 10.5.3, c. See also A. Kotaite, "Investment and Training Needs Among the Challenges Facing Developing Countries" (1993) 48:2 *ICAO Journal* 24 at 26.

85 See WW/IMP-WP/13, *supra* note 81 at para.3.1.6.

86 See Vidler, *supra* note 77 at 73. In his words, "airlines will be ultimately presented with the bill tomorrow for today's haste."

87 See S.A.Kaiser, "Infrastructure, Airspace and Automation – Air Navigation Issues for the 21st Century" XX:1 (1995) *Annals of Air and Space Law* 447 at 453.

PART II

INSTITUTIONAL ASPECTS

2 The Evolving GNSS

Signal Providers: Characteristics and Policy Issues

Global Positioning System (GPS)

The origins of the GPS can be traced back to the early seventies, when research for an U.S. defence navigation satellite system led to the development of the Navstar Global Positioning System.[1] Conceived by the U.S. Department of Defence (DOD) to enable positioning of military equipment, including land vehicles, ships, aircraft and precision-guided weapons anywhere in the world, providing global coverage with a ten-metre accuracy,[2] the GPS was deployed over two decades at a cost of U.S. $ 10 billion. Having proven to have excellent capabilities in its defence role, it has been integrated into virtually every facet of U.S. and allied military operations,[3] which are increasingly reliant on its signals for a variety of purposes, from navigation to modern precision-guided weapons and munitions.[4]

However, whereas the system brings about countless benefits to the U.S. armed forces, it obviously carries multiple countervailing risks, so that the more dependant the military become on GPS, the more vulnerable they are to potential signal disruptions. Even though eventual nuclear adversaries might not need GPS-level accuracies to cause significant damage, hostile exploitation of GPS is possible and GPS-aided weapons may pose a significant threat if they manage to evade U.S. defence. To cope with future threats, the DOD must develop selective denial techniques, such as tactical jammers to deny positioning and navigation from GPS and differential GPS-based systems, as well as defence programmes against cruise missiles and ballistic missiles that

37

may carry conventional warheads or weapons of mass destruction. Under this approach, it will be possible to effectively deny GPS signals to adversaries in conflict areas while preserving its peaceful use in the rest of the world. It has been argued that it is in the security interests of the U.S. to have differential GPS networks outside national boundaries controlled by allied nations, as opposed to potential adversaries or international organisations. Direct control could encompass a variety of techniques, ranging from encryption of communication links to diplomatic agreements that would limit areas and times of operation when circumstances warrant.[5]

It was in the wake of the Korean Airlines disaster[6] that President Reagan declared that GPS would be offered for free[7] to the civilian community.[8] From then on, GPS has evolved far beyond its military roots and rapidly emerged into public awareness.[9] It has now become vital to telecommunication and transportation infrastructures, including air, surface, marine and rail navigation, and supports, on a global basis, a wide range of civil, scientific and commercial activities. Potential uses apart from civil aviation include, *inter alia*, car navigation, fleet management, delivery services, motorway/waterway maintenance, search and rescue, roadside assistance and other emergence responses, mineral and resource exploration, natural resource management, farming, wildlife tracking, recreation activities (hiking, camping, hunting, boating, and fishing), satellite tracking and data processing, future space station operations, weather forecasts, mapping, surveying, and other time-critical applications.[10]

Nevertheless, many current applications were not considered at the time of the original planning and configuration of the systems. For this reason, there has been some concern to investigate and address, in particular, the adequacy of the current GPS configuration, its future capabilities and dual (military/civil) character.[11] As a result of a comprehensive policy review jointly conducted by the White House Office of Science and Technology Policy and the National Security Council, a "U.S. Global Positioning Policy"[12] was announced on the 29th of March, 1996 by Vice-President Gore, presenting a strategic vision for the future management and use of GPS. Other studies have been undertaken, such as the assessments made by the National Research Council and the National Academy of Public Administration,[13] and the Rand Corporation.[14]

The dual-mode use of GPS is a particularly relevant issue, since it makes GPS both a domestic asset and an international resource at the same time. The mechanism can be explained as follows. GPS satellites transmit two different signals, namely, the Precision P-code and the Coarse Acquisition or C/A code. Providing what is called the Precise Positioning Service (PPS), the P-code is designed for authorised military use only,[15] and is available on both L1

(1575.42 MHz) and L2 (1227.6 MHz) frequencies. It provides the highest accuracy services as regards positioning, velocity and timing information available on a continuous, world-wide basis to authorised users, that is to say "a positioning accuracy of at least 22 metres horizontally and 27.7 metres vertically, and time transfer accuracy to co-ordinated Universal Timing (UTC) within 200 nanoseconds".[16] An encryption process such as anti-spoofing (AS) can be installed so as to prevent acquisition by unauthorised users.[17]

The C/A code provides the Standard Positioning Service (SPS) for use by non-military users. Less accurate than the P-code, and consequently more prone to jamming, It is available on the L1 frequency and provides positioning accuracy of 100 metres horizontally and 156 metres vertically, and time transfer accuracy to UTC within 340 nanoseconds, with a probability of 95 percent. For national security reasons, its accuracy was intentionally degraded by the U.S. military by imposing the so-called selective availability (SA) on all GPS Block II satellites as of March 25, 1900.[18]

Throughout the years, advisory committees strongly recommended the removal of the SA in peacetime, arguing that the risk of encouraging GPS-aided weapons should be balanced against the benefits of using GPS for satellite-based navigation.[19] The Presidential Decision Directive of 1996 revealed the intention to discontinue the use of GPS Selective Availability within a decade. Initial consideration for its removal would begin in 2000, hence the President would make an annual determination on the issue, in co-operation with the Secretary of Transportation, the Director of Central Intelligence, and heads of other appropriate departments and agencies.[20] After careful examination, President Clinton finally announced the U.S.' decision to discontinue the use of SA on the 1st of May, 2000.[21] The measure has since dramatically improved the accuracy of the GPS signal available to the public from hundreds of metres to about twenty metres, therefore greatly enhancing the SPS for all civilian applications, and further allowing for GPS's expanded civil use and accelerated acceptance world-wide. Yet, as far as aviation is concerned, it is important to note that the ensuing levels of services are still not capable of meeting the strict RNP requirements for precision approaches. Differential corrections, which are normally provided to the individual GPS user by means of augmentation techniques, increasing accuracy to 5 metres and in some cases even to the sub-metre level, as well as warnings of system interruptions or interference, are still necessary to guarantee additional accuracy, system availability, integrity, and continuity of services for safety-of-life operations.[22]

Driven by the need of system architectural improvements, including the overall system functions and configuration, details of signal structure, augmentations and constellation enhancements, various studies[23] were recently completed within the U.S. government. Apart from the termination of SA, an increase in the size of the GPS constellation to 30 - 36 satellites is being considered as part of the future evolution of the system. A second coded civil frequency for general use in non safety-critical applications will be added to L2, at 1227.60 MHz, on Block IIF satellites scheduled to be launched in 2003. Addition of a third civil signal (L5) has also been considered so as to improve redundancy and enhance the capability of GPS. The signal would be provided at 1176.45 MHz on Block IIF satellites to be launched in 2005. By contrast with the new signal on L2, this signal would be available for critical safety-of-life applications due to its location in the protected ARNS/RNSS frequency band. These additional frequencies will also be free of direct user charges and are expected to be available for initial operational capability in approximately 2010.[24]

When the need arises to balance national security, foreign policy and economic interests, due consideration must be given to the fact that whilst GPS's space and control segments are under U.S. jurisdiction, the user segment is already in the hands of the private sector all over the world. Thus, the reason for strictly national control which emerges from the guidelines of the GPS Policy Statement as an overriding priority. Although encouraging private sector investment in and use of U.S. GPS technologies and services, and promoting international co-operation in its use for peaceful purposes, the Department of Defence will continue to acquire, operate, and maintain the basic GPS.[25] Furthermore, the government is committed to preserving and upgrading the full military utility of GPS, having already demonstrated the capability to selectively deny GPS signals on a regional basis when national security is threatened.[26]

As for the management and operation of GPS and U.S. government augmentations, according to the Policy Statement, GPS has progressed from exclusively military control to be managed by a permanent interagency GPS Executive Board, jointly chaired by the Departments of Defence and Transportation, the latter being responsible for all federal civil GPS matters.[27] Commenting on the issue, Dr. Wulf v. Kries has stated that:

> Nowhere in the policy statement foreign participation is considered as extending to the system's governance. ... Institutionally, therefore, GPS will remain a military system. As such it is not suitable for internationalisation. ... Any form of international participation would be conditioned on compliance with U.S. military requirements. This inevitably rules out a multi-partite GPS partnership.[28]

Moreover, a reading of the Statement clearly indicates that the U.S. government is determined to institute GPS as an undisputed global monopoly: by "continu[ing] to provide the GPS Standard Positioning Service ... on a continuous, world-wide basis, free of direct user fees",[29] the U.S. plans to enhance the already stimulated growth of commercial GPS applications. This fact coupled with the discontinuance of the selective availability is even expected to serve as a deterrent to international competition. Such ambition becomes manifest when the government advocates the acceptance of GPS and U.S. government's augmentations as standards for international use.[30]

Reality, however, shows otherwise. The RAND Study reminds that competition with GPS is a possibility, which could endanger the U.S. lead in satellite navigation technology and commercial exploitation, especially if the U.S. were to fail to sustain the GPS constellation or to provide competent and reliable services, or were to charge users for access to the signal, thus unintentionally creating an inviting opportunity for a competing system.[31]

Furthermore, the international community is preparing itself for the development of a civilian-controlled GNSS. The European Commission has set out a strategy to secure a full role for Europe in the development of the next generation GNSS. It follows that it has proposed to develop "an integrated European GNSS system, Galileo, open to other international partners, that is independent of the GPS system but complementary to and fully interoperable with GPS".[32] Detailed discussion on the subject will follow below.

Global Navigation Satellite System (GLONASS)

From its inception, in the middle of the 1970s, the GLONASS medium-orbit global navigation satellite system was developed by the former Soviet Union as a dual-purpose system for both defence needs and for civil use, much like its American counterpart. However, the systems differ on three important aspects:

Firstly, GLONASS orbits have a greater inclination than those of GPS, which makes it possible to have a greater number of satellites simultaneously visible to users in the middle and high altitudes. Secondly, while GPS satellites use but a single frequency to operate on, and signal division is performed by the "code method", GLONASS uses frequency division of its signals, each satellite transmitting the same code at a different carrier frequency. As a result, the systems' immunity to interference is much strengthened, although a wider frequency spectrum becomes a necessary condition. Thirdly, and most significant of all, so far the Russian Federation has never demonstrated any intention of degrading the channel intended for civil use. Suffice it here to say that the actual position-finding accuracies in GLONASS are 57-70 metres horizontally and 70 metres vertically, with a probability of 99.7 percent.[33]

Radionavigation signals are presently transmitted in two frequency bands, namely L1 (1597-1617 MHz) and L2 (1240-1260MHz). Each satellite transmits two codes: the so-called Channel of Standard Accuracy, designed for civil use, on the L1 frequency, and a high precision P-code on both L1 and L2 frequencies. Future evolution of the basic system includes its transfer to civil control and its promotion to benefit such users. Thus, addition of a new ranging signal on L2 for civil use is under consideration by the Russian authorities. The overall objective in the planned evolution is to improve performance characteristics and enhance its capabilities as one of the GNSS elements. The next generation known as GLONASS-M may include: i) enhanced data structure allowing better combined use of GLONASS and GPS; ii) modernised space segment; iii) additional signal power on L2; iv) provision of P-code on L1/L2.[34]

Furthermore, on February 18, 1999, a decree by President Yeltsin created a joint-military-civilian board to operate GLONASS, reaffirming the dual purpose of the system. The need for funding along with the access to highly valuable radio frequencies has determined a change in policy. Besides allowing for private foreign investment, the system has been offered as a basis for an international global navigation satellite system.[35] Thus the possibility exists for co-operation between the European Commission and the Russian Federation, so that GLONASS would participate in the development of a European Global Navigation Satellite System.

Satellite-based Augmentation Systems

Wide Area Augmentation System (WAAS)

The WAAS is being developed by the FAA (U.S.) to satisfy requirements of primary means navigation down to, and including Category 1 precision approaches, which are not met by the basic GPS service. It is also expected to improve system accuracy to 7.6 metres vertically and horizontally, and to provide increased availability and integrity information about the entire GPS constellation.[36]

In its initial phase of implementation, it consists of an integrated network of ground reference stations which receive GPS signals and, having determined if any errors exist, relay this data through terrestrial or satellite communication links to the wide area master station where correction information is computed and uplinked to INMARSAT-III geostationary communication satellites. The message is then broadcast on the same frequency of GPS to receivers on board the aircraft flying within the coverage area.[37]

Local Area Augmentation System (LAAS)

Intended as a complement to the WAAS, the LAAS will be used where the former is unable to meet existing RNP requirements. For instance, it may be used to provide Category I precision approach capabilities in areas with extremely high availability requirements or where GPS signal reception is inhibited by geographical limitations. In addition, it will provide the extremely high accuracy necessary for Categories IIIa and IIIb precision approaches and is expected to be able to pinpoint an aircraft's position to within one metre or less.[38]

Multi-functional Transport Satellite Augmentation System (MSAS)

Responsible for the provision of air traffic services in the Tokyo and Naha Flight Information Regions (FIRs), which covers a vast congested area connecting the Asia/Pacific region to North America, Japan saw the need to develop the Multifunctional Transport Satellite (MTSAT) to cope with the rapid increase in traffic and to ensure safety and efficiency in the implementation of the CNS/ATM systems.[39]

The MTSAT, currently under implementation, is composed of geostationary satellites with aeronautical and meteorological payloads, the former providing communication and navigation functions.[40] Its navigation component, the so-called MSAS, provides three types of GNSS augmentation information, namely, ranging, integrity and differential information.[41]

Policy considerations indicate that the Japan Civil Aviation Bureau will continue to offer MTSAT for use as a common infrastructure in the Asia/Pacific region on a not-for-profit basis. Its use is not mandatory, even within Japanese FIRs. As seamlessness and interoperability with other augmentation systems are desired, co-operation and close work with related organisations is necessary and is being sought.[42]

European Geostationary Navigation Overlay System (EGNOS)

In December 1994, Europe finally came out with its own strategy as regards satellite navigation. The Council of the European Commission welcomed then, in a resolution, the Commission's proposal "to initiate or support work needed for the design and organisation of a global navigation satellite system for civil use".[43]

The European Tripartite Group (ETG)[44] was subsequently set up to facilitate a harmonised and concerted European contribution to the next generation of global navigation satellite systems.[45] Reflecting the multimodal and international nature of the systems, the group is composed of the Commission of the European Union, Eurocontrol and the European Space Agency (ESA).[46]

The approach being taken by the group comprises two steps. Firstly, "to make the best and earliest possible use of systems based on GPS and GLONASS", by engaging in the provision of augmentation services. Secondly, and meanwhile, "to develop and deploy an independent civil successor which will not suffer from the technical and institutional limitations of the current systems",[47] "and at the same time will facilitate access for European industry into the global market for systems and services".[48] To distinguish these two phases, they are referred to, respectively, as GNSS-1 and GNSS-2.[49] EGNOS is the European contribution with respect to GNSS-1; Galileo, the proposed development for GNSS-2.

EGNOS' space segment will use the transponders on the Inmarsat-III Atlantic Ocean Region East AOR(E) and Indian Ocean Region (IOR), and Artemis geostationary satellites to perform its augmentation functions, namely geostationary ranging, integrity monitoring and wide area differential, and is expected to achieve full operational capability in 2002. It will be fully interoperable with the other satellite-based augmentation systems.[50]

As for the ground segment, ranging and integrity monitoring stations equipped with GPS/GLONASS/GEO receivers, atomic clock and weather sensors act as data collectors. Such data is transmitted to master control centres which estimate errors and produce the necessary augmentation messages then relayed to navigation land earth stations, where a GPS-like navigation signal is generated and uplinked to GEO satellites, and, through these, transmitted to users. There are two land earth stations per GEO satellite, one active and one in hot backup. The communication network of the EGNOS ground segment is based on terrestrial and satellite links, depending on the geographical location and service availability, but a fully redundant network is to be deployed.[51]

Emerging GNSS Elements

Galileo

The proposed development of a constellation of satellites, Galileo, intended to become a key element of the Trans-European Network (TEN) for navigation, timing and positioning services, and a new element of GNSS, is part of the strategy set up by the European Union, in collaboration with the European Space Agency, to secure a full role for Europe in the development of the next generation of Global Navigation Satellite Systems.[52]

Lack of European influence on a system which is increasingly gaining importance in nearly all fields of technology and becoming central to all forms of transport would place the industry in an extremely disadvantageous position and seriously constrain its capacity to compete in the rapidly expanding markets. Should the situation remain as it is, European users would be left entirely dependent on a foreign system for the provision of services.[53] In fact, the European Commission particularly advocates that the absence of any guarantees of service on the part of the U.S. coupled with military control of GPS management shall definitely accentuate Europe's vulnerability to general loss or selective denial of services in the event of war.[54] In addition, it fears the future imposition of "unilaterally decided and excessive charges", when there would only be "a limited possibility of quickly developing alternatives".[55]

Recalling the existence of important industrial, strategic, military and political interests for Europe in the control of the systems, in January 1998, a Commission Communication[56] proposed an approach involving the development, at the European level, of a system which would fully meet the requirements for its civil use.

With a view to providing Europe with the capability to deliver a global service that would meet the requirements of safety-critical applications, the Commission envisaged the development of an integrated European system, Galileo, global in coverage from the outset, open to all international partners, technically independent from the GPS system but complementary to and fully interoperable with it, and which would exploit new state-of-the-art capabilities in an autonomous civil system. The main objective therein would be to make the overall GNSS robust, and yet remedy certain shortcomings of GPS with the introduction of a guaranteed service for integrity monitoring by means of a dysfunction warning system.[57]

The overall approach presented by the Commission was approved by the European Council in June, 1999.[58]

The development of the basic navigation system shall encompass four phases, namely a definition phase (concluded in 2000), a development and validation phase (2001-05), a deployment phase (2006-07) and an operational phase thereafter.[59] It will be based on a core constellation of 30 medium-earth orbit satellites,[60] combined with the appropriate infrastructure of global, regional, local and user components, and with a currently estimated cost of approximately EUR 3 billion.[61]

A resolution[62] followed on 19 July 1999, whereby the Council invited the Commission, *inter alia*:

i. to fully explore possibilities for co-operation and/or future development with the U.S. and the Russian Federation while continuing technical consultations;
ii. to explore the interest of other third countries to co-operate in the area; ...
iii. to present a thorough cost-benefit analysis encompassing all relevant options for the whole project, and within this framework to:
    a) examine scenarios for the creation of revenue sources ...;
    b) develop and to present at the beginning of the year 2000, framework conditions for the proposed public-private partnership, including an appropriate distribution of roles and tasks, as well as costs and risks ...;
    c) create timely and realistic conditions for securing finance largely from the private sector ...
iv. ...to start without delay in co-operation with the ESA and the Member States, the definition phase of the project ...[63]

Accordingly, the possibility of international co-operation was studied and it was recognised that joint development of the next generation GNSS was likely to be the most cost-effective alternative, as long as certain conditions were satisfied, namely: i) full European participation in the future design, development and operation of GNSS; ii) firm guarantees against unilateral suspension of services; and iii) an opportunity for the European industry to compete in all segments of the related market.[64]

Whereas the U.S. is not willing to share control of GPS for the reasons already expressed, negotiations are in progress to ensure compatibility and interoperability between both systems. With a view to reducing investment costs, the Russian Federation is effectively offering full co-operation in the development of a new international civil system, based on the coexistence of both Galileo and GLONASS constellations.[65] The principal advantages of this approach would be the shared use of the valuable GLONASS frequency allocation[66] as well as Russian know-how in satellite operation and control, which would allow for a rapid development of the systems. Co-operation with Canada, Israel and other third countries is currently being investigated in regard to areas of particular national interest, ranging from direct investment or industrial co-operation to user applications and equipment.[67]

As a result of large-scale consultations with potential users, the approach favoured is to develop three different levels of service in order to fully meet users' requirements. At a first level (General Service), a basic Galileo signal for mass-market applications with no special needs for guarantee of service or integrity information, and to which there would be universal access, would be provided free of charge, in consistence with the present U.S. policy. At a second level (Commercial Service), a fully certifiable controlled access service would be available to subscribers subject to user charges. Here, two categories have been identified, namely an Accuracy and Integrity Service for less demanding safety of life users and professional markets, and a precise Ranging and Timing Service, intended, *inter alia*, for surveying, meteorological forecasting, etc. Finally, at a third level, a High Integrity Service (Public Service) would be provided which, depending on further decisions to be taken by the European Union or international organisations, might be restricted to authorised users by means of signal encryption. With the highest guaranteed availability, system integrity and accuracy, it is expected to satisfy institutional commitments and international requirements for safety-of-life, governmental applications and security-related services, such as civil aviation, emergency services, road tolls, as well as search and rescue applications.[68]

It is irrefutable, however, that the provision of different levels of services with various features will require Community regulation and standardisation as well as an enforcement mechanism allowing for the monitoring of compliance by the competent bodies.[69]

In addition thereto, security constraints with regard to protection against interference and misuse have clear implications for system design and need to be resolved before the development and validation phase can begin, so as to ensure physical protection of the system and allow for controlled access to certain services.[70]

Other key decisions in the definition phase respect to the gradual integration of EGNOS into the architecture and structure of Galileo, expected to be completed by 2008.

Recognising the difficulties in generating revenue from Galileo whilst the GPS SPS is provided for civil use free of charge, ruling out the possibility of Galileo being provided exclusively by the private sector, the Commission suggested that it be developed as a public-private partnership (PPP), and proposed the following financing strategy:

a) substantial financing at European level, through the EU budget, notably the TEN, research and development programmes and, the ESA;

b) establishment of revenue streams, which is likely to require regulatory action;
c) developing a public-private partnership to deliver complementary finance and value money and to ensure that user's needs are met.[71]

Whereas cost/benefit studies have shown Galileo to be cost-effective, and therefore sufficiently attractive to private investment, it is undeniable that substantial public funding will be essential to ensure the Community's political control over the programme as a whole. In fact, the estimated costs for the next phase (development and validation) will be entirely met from European public resources already envisaged for the current financial period, namely European Union (TEN, 5th Framework research programme, or other future research actions) and ESA funds. No funding in the form of subsidies is expected after the completion of deployment phase.[72]

Suggested sources of revenue which may also be mobilised include, *inter alia*: i) introduction of levies on receivers for all satellite-based applications; ii) licence fees for services suppliers and operators; iii) provision of controlled access services against fees; iv) contributions from non-member States using the system; and v) mandatory use of certain services by means of public regulation.[73]

On the other hand, the priority for the Galileo project definition phase has been to fully explore the overall market possibilities so that the necessary political and financial decisions can be taken with a view to mobilising the private sector to make the necessary financial commitments on the basis of future expected economic revenue.[74]

Private sector involvement in the Galileo programme from an early stage is perfectly justifiable in the need of effective management and optimum cost-control for the sound commercial operation of the services to be provided. Possible forms of contribution range from direct financing to the development of equipment or transfer of technology, to cite just a few.[75]

In this regard, the establishment of a public-private with the creation of a vehicle company, accountable for project delivery but with efficiency guaranteed by management autonomy, might prove to be an essential feature in order to reassure potential investors of the financial stability of the programme.[76]

It is to be put in place through public tender by a Public Management Board established for these purposes. The Board is to be succeeded in the operational phase by a Galileo Administration which will be responsible for managing operations, guaranteeing performance and security co-ordination, while contracting out the actual operation of the system to the European GNSS service provider (the vehicle company). In turn, the service provider might choose to directly provide services to the end users or either to delegate their provision to another third party service provider.[77]

With a legal personality, the Administration will be responsible for responding to any liability claim relating to Galileo.[78] Whether and to what extent there may be a partial or total "release of liability" of a State when outsourcing the provision of GNSS signals, services and facilities to a foreign entity will be subject of detailed consideration in Chapter 3.

It is clear that the main benefits of Galileo are political rather than economic, particularly the advantage of retaining control over safety critical services.[79] Nonetheless, it is still necessary to distinguish between the social-economical desirability and the financial viability of the project, since most expected benefits will not result in revenue without regulatory action at the public level. Therefore it is the endorsement of a PPP approach, whereby users' requirements will have a central role, the means to help improve value for money and make the private sector confirm its commitment to the project by investing risk capital in it.[80]

In any event, the challenge is to act decisively and in time. Otherwise, the planned evolution of the GPS will reinforce its market dominance in a way that it may finally be adopted as "the" standard, leaving Europe to play but only a mere supporting role.

The Way Forward

Following the Galileo experience and other recent developments, including the proposed GPS L5 signal, it has been acknowledged that the evolution of GNSS will be an incremental process, "which cannot and should not be bound to predefined GNSS configurations".[81] The initial package of SARPs recommended by GNSSP/3 has been developed so as to be able to easily accommodate any modifications or new elements.

The optimum design architecture of any future navigation system, the so-called long-term GNSS or GNSS-2, will have to satisfy many user applications apart from civil aviation. Each will require a different level of safety and accuracy performance. It must be need-driven to be commercially attractive and financially justifiable. In particular, it should evolve from the existing elements, maintaining full interoperability with them in order to enable a timely and cost-effective transition.

Finally, whereas many political obstacles indeed have to be overcome, and an agreement still must be reached on many legal, financial and institutional issues, the security offered by an internationally-controlled civil system promises to be the best solution for the long-term.[82]

Evolutionary Introduction of GNSS

GNSS as a Sole-Means Navigation System

In accordance with the Global Plan, GNSS implementation must be carried out in an evolutionary manner, allowing for system improvements to be gradually introduced.[83] Guidelines for transition to the future systems encourage the earliest possible accrual of its benefits starting with supplemental en-route use. Three levels[84] have been identified by the FANS Phase II for introduction of GNSS-based operations, namely the use of GNSS as a supplemental means of navigation, as a primary means, and as a sole means of navigation, the latter representing the ultimate goal, when GNSS must allow aircraft with the required state of avionics equipment to meet all four RNP requirements for a given operation or phase of flight.

The terminology above is a further indication that operational approvals for aircraft must be issued for particular operations and should identify specific conditions or restrictions to be applied. To this end they might vary by States.[85]

Whether GNSS will become the sole navigation system of the future has been widely discussed world-wide.[86] Although the ground infrastructure of the current navigation systems must remain available during the transition period to ensure the reliability of the new system, it is undeniable that considerable savings would be accrued if such technology could be put aside in exchange for sole dependence on GNSS. However, opinions seem to differ as to the safety there really is in relying on the sole means GNSS. While some have stated that "[t]here is safety in the existence and availability of several GNSS systems, each of which can provide back-up in case another system goes down",[87] others have argued, for example, that the U.S. will never rely solely on satellites for its self-defence to the point of turning off its ground-based air navigation system.[88]

There are, on the other hand, a number of factors that might influence the performance of GNSS, all of which raise important concerns with respect to the sole reliance on the services provided, as follows.

Discontinuation of Services, Unlawful Interference and Other Concerns

Apart from the intentional degrading of the civil signal, which has been discussed above, these concerns include the unilateral suppression of GNSS service in a conflict zone, or else the disruption of the signal by hostile military forces. Legal considerations have indicated that selective denial of signals to such users for national security purposes might not be considered illegal after all in such circumstances. In this sense, the Chicago Convention clearly stipulates that, in case of war, the freedom of action of any of the contracting

States affected, whether as belligerents or as neutrals, shall not be affected by its provisions.[89]

Furthermore, international law recognises the well-known roman maxim *salus populi suprema lex* as a correlative to the fundamental principle of "self-preservation", whereby the State may take all necessary measures to protect the nation against external danger and hostility. Under certain circumstances, the State may "even disregard a minor right of another State or its nationals in order to preserve its own existence".[90] In the words of Bin Cheng:

> In the present structure of international society, however, where a State stands on the one hand as a supreme political institution, sovereign within its boundaries, and on the other hand as a member of a society in which other equally sovereign members co-exist, ... a proper knowledge of the limits and conditions of its application is just as important as knowledge of the existence of the principle itself. ... The only legal limitation on the discretion of the States appears to be the principle of good faith. The measures taken should be reasonable and must not be arbitrary, oppressive or maintained longer than necessary.[91]

B.D.Henaku, invoking the doctrine *sic utere tuo ut alienum non laedas,* argues that the signal provider State is under no obligation to provide the signal, but where it opts to do so, its sovereign rights are curtailed and the State is "restrained from taking actions that could cause transnational damage".[92]

Principal among security concerns therefore is the need to protect access to the signal for safety-critical uses from potential threats of intrusion, unlawful interference or jamming, that might disrupt GNSS services over relatively large areas. States and service providers should conduct investigations on the improvement of techniques to prevent or minimise the effects of jamming and spoofing.[93]

Further consideration should be given to the existing international and domestic law providing sanctions for unlawful interference with air navigation aids so as to determine whether they are reasonably adequate to cover foreseeable threats in respect to the CNS/ATM systems.

Preliminary observations with respect to the 1971 Montreal Convention[94] and the 1963 Tokyo Convention[95] indicate that both instruments only apply to unlawful acts committed with specific intent to cause damage. Thus, it appears that the law is silent on the issue of interference with data through "hacking", where the offender lacks the required degree of intent. In addition thereto, the Conventions specifically refer to damages to "air navigation facilities". It is not yet clear whether the primary GNSS signals will be considered to fall into the scope of these provisions. The national laws implementing the referred clauses in the States must be examined so as to determine whether the new CNS/ATM technology is contemplated therein.[96]

Other concerns can be cited. Some are environmental-related, such as ionospheric activity which, following an 11-year solar cycle, may constitute the largest error component in GNSS, requiring continuous differential correction.[97] Furthermore, interruption of the services due to budgetary constraints cannot be forgotten as another contingency. The American offer and its Russian counterpart have actually been made subject to, respectively, the availability of government funds and allocation of resources.[98]

The potential of GNSS to provide seamless navigation guidance globally and for all phases of flight has been promoted by ICAO. The ability of the system to do so is not yet fully demonstrated, and a number of concerns are still being addressed:

> To date, no evidence is available to conclude that this potential cannot be realised. Shortcomings of today's GNSS are being mitigated through the introduction of augmentations and integrated applications of these augmentations, and other limitations can be overcome by evolutionary development towards the long-term GNSS.[99]

In brief, GNSS sole means approval is therefore a necessary, but not sufficient condition for termination of present radio navigation services.[100] An eventual and progressive withdrawal of current radio navigation systems will depend on many factors, among which are the implementation and quality of the new systems and their robustness. It will definitely take into consideration safety and cost-benefit studies, as well as the progress of regional and global co-ordination through ICAO. Thus the actual transition will be largely determined by the degree of confidence in the performance of GNSS, and will probably differ in various regions of the world.[101]

Notes

1   See K. D. McDonald, "Technology, Implementation and Policy Issues for the Modernisation of GPS and its Role in a GNSS" (1998) 51:3 *The Journal of Navigation* 281 at 281-282 [hereinafter McDonald].

2   See P. A. Salin, "Regulatory Aspects of Future Satellite Air Navigation Systems (FANS) on ICAO's 50th Birthday" (1995) 44:2 *Zeitschrift sur Luftund Weltraumrecht* 172 at 172 [hereinafter Salin].

3   Not only is GPS used by the U.S. military, but also by foreign military in NATO, who are expected to continue to use the system, in order not to raise co-ordination issues within the organisation. See P. B. Larsen, "Future GNSS Legal Issues" (Third United Nations Conference on the Peaceful Uses of Outer Space, UNISPACE III, 19-30 July 1999) at 6.

4   See W. v. Kries, "Some Comments on U.S. Global Positioning System Policy" (1996) 45:4 *Zeitschrift sur Luftund Weltraumrecht* 407 at 407 [hereinafter Kries].

5   See S.Pace, "The Global Positioning System: Policy Issues for an Information Technology" (1996) 12:4 *Space Policy* 265 at 267-268 [hereinafter Pace].

6   Korean Airlines Flight KE007, from New York to Seoul, South Korea. On 1 September 1983, the aircraft strayed into Soviet airspace and was shot down by Soviet military aircraft over the Sea of Japan. All 269 persons aboard were killed. For related Court decisions, see Bowden v. Korean Air Lines, 814 F. Supp.592 (E. D. Mich., 1993); In re Korean Air Lines Disaster of Sept. 1, 1983, 807 F. Supp.1073 (S.D.N.Y. 1992). Compare Park v. Korean Air Lines, 24 Av. Cas.(CCH) 17,253 (S.D.N.Y. 1992). See also S. Kaiser, "A New Aspect of Future Air Navigation Systems: How Secondary Surveillance Radar Mode S Could Protect Civil Aviation" (1992) 41:2 *Zeitschrift sur Luftund Weltraumrecht* at 154-164.

7   General tax revenues (DOD and U.S. Coast Guards costs) and air transportation trust funds, supported by a fuel tax or value added tax, are currently used to finance the costs of GPS services. See Larsen, *supra* note 3 at 9.

8   See Salin, *supra* note 2.

9   Over one million GPS receivers are now produced annually. Projections are for GPS to be a 31 billion-dollar market by 2005.

10   See U.S., *National Civilian GPS Services* (Washington, D.C., Department of Transportation, 21 March 2000) at 12.; U.S., *Civilian Benefits of Discontinuing Selective Availability – Fact Sheet* (Department of Commerce, 1 May 2000); Pace, *supra* note 5 at 265-266.

11   See McDonald; *supra* note 1 at 281-282.

12   See U.S., *U.S. Global Positioning Policy* (The White House Office of Science and Technology Policy and the National Security Council, 29 March 1996) [hereinafter GPS Policy Statement or Presidential Decision Directive].

13   See U.S., *The Global Positioning System. Charting the Future* (Washington, D.C., National Academy of Public Administration and National Research Council, 1995) (Chair: J.R. Schlesinger) [hereinafter NAPA/NRC Report].

14   See RAND Critical Technology Institute, *Global Positioning System. Assessing National Policies* (Santa Monica: Rand, 1995) (Dir.: S. Pace) [hereinafter RAND Study].

15   PPS is available to U.S. and allied military users. Limited access for civilian use may be available upon request through special agreement with the DOD. See U.S., *Global Positioning System Data and Information Files* (U.S. Naval Observatory, Automated Data Service), http://tycho.usno.navy.mil/gps.html (date accessed 12 August 2000) at 1 [hereinafter USNO Data Service].

16   Larsen, *supra* note 3 at 9.

17   See Pace, *supra* note 5 at 266-267.

18   See USNO Data Service, *supra* note 15 at 1-3; McDonald, *supra* note 1 at 269.

19   See Pace, *supra* note 5 at 269.

20   See Presidential Decision Directive, *supra* note 12 at paras.III (2) and V.

21   See U.S., *Statement by the President Regarding the United State's Decision to Stop Degrading Global Positioning System Accuracy* (White House, 1 May 2000) [hereinafter *President Statement*].

22   See ICAO, *9th Meeting of the Caribbean and South American Regional Planning and Implementation Group*, GREPECAS/9 (7-12 August 2000), "Recent Developments in the Modernisation of the Global Positioning System and U.S. Satellite Navigation

Program Status", presented by the U.S. [hereinafter GREPECAS] at paras.3–4. See also McDonald, *supra* note 1 at 289.

23 The DOD through the GPS Joint Programme Office (GPO) has analysed the feasibility of a wide range of future architectural options, the results of the study having been published in the Acquisition Master Plan in 1997. So have the Defence Science Board, the USAF Scientific Advisory Board and the NRC, having completed their investigation in 1995. See McDonald, *ibid.* at 285.

24 See ICAO, *Report on the Third Meeting of the Global Navigation Satellite System Panel*, GNSS/3-WP/66 (12-23 April 1999) at para.3.2.1 a, b. [unpublished][hereinafter *GNSSP Report*]. See also, *ibid. U.S. Rationale for the Selection of GPS L5*, Appendix E to the Report on Agenda Item 1.

25 See Presidential Decision Directive, *supra* note 12 at IV (1). See also Kries, *supra* note 4 at 408.

26 See *President Statement, supra* note 21.

27 Presidential Decision Directive *supra* note 12 at III (7), IV (1).

28 See Kries, *supra* note 4 at 408.

29 Presidential Decision Directive, *supra* note 12 at III (1). The U.S. has pledged before the General Assembly of ICAO to continue to provide GPS signals free of direct user charges to the international civil aviation community. This pledge was affirmed in the Directive issued in 1996, and the commitment was solidified by an Act of Congress in 1997, which established as a matter of law the provision of GPS services for peaceful, civil, commercial, and scientific uses on a continuous world-wide basis free of direct user charges. Moreover, in his recent Statement of May 1, 2000, President Clinton reaffirmed the government's commitment to provide free and improved GPS capabilities to world-wide users free of charge. See ICAO, *156th Session of the Council, Policy on the Future Use of the Global Positioning System*, ICAO C-WP/11097 (9 March 1999), presented by the U.S. of America at 2; *President Statement, supra* note 21. But see L. Bond, "The GNSS Safety and Sovereignty Convention of 2000AD" (Global Airspace 99, Washington DC, 3 February 1999) [unpublished], where the speaker highlights the fact that the language of the PDD is not clear and so deserves careful reading. In his view, there are two critical conflicting clauses: "[t]he first says that GPS will be provided continuously, without charge, for civil purposes. The second says that GPS will remain responsive to the National Command Authority, i.e. the President of the U.S. So the PDD reserves to the U.S. the right to turn off, degrade, or spoof GPS whenever it wants without prior notice or explanation".

30 See Kries, *supra* note 4 at 409.

31 See Pace, *supra* note 5 at 270.

32 EU, *Communication COM (1999) 54 final of 10 February 1999, Galileo, Involving Europe in a New Generation of Satellite Navigation Services* [1999] Bulletin EU 1/2 1999, Transport (5/23) [hereinafter *COM (1999) 54 final*] at 1.3.169.

33 See V. Kuranov and Y. Iovenko, "Capability and Performance Make GLONASS Suitable for Navigation in All Phases of Flight" (1997) 52:9 *ICAO Journal* 11 at 11.

34 See *GNSSP Report, supra* note 24, *Appendix to the Report on Agenda Item 3* at 3.2.2.

35 See Russian Federation, *Directive of the President of the Russian Federation No. 38-rp* (18 February 1999), http://mx.iki.rssi.ru/SFCSIC/english.html (date accessed: 10 August 2000) [*hereinafter Directive of the Russian Federation*].

36 See J.C.Johns, "Enhanced Capability of GPS and Its Augmentation Systems Meets Navigation Needs of the 21st Century" (1997) 52:9 *ICAO Journal* 7 at 7. Preliminary tests indicate that total system accuracy is approximately 2-3 metres (horizontal and vertical). See *GREPECAS, supra* note 22 at para.5.

37 See *GREPECAS, ibid.*

38 See J.C.Johns, "Navigating the 21st Century with GPS" (1997) 39:3 *Journal of Air Traffic Control* 34 at 34-35.

39 See K. Fukumoto and K. Abe, "MTSAT: Japanese Contribution to the Implementation of ICAO CNS/ATM Systems in the Asia/Pacific Region" (1998) 46:184 *Revue Navigation* 442 at 443.

40 See WW/IMP, "MTSAT: Japan's Contribution to the Implementation of the ICAO CNS/ATM Systems in the Asia/ Pacific Regions", ICAO WW/IMP-WP/45 (11 May 1998) [hereinafter WW/IMP-WP/45] at 2.

41 See K.Fukumoto and K.Abe, "First of Several Japanese Satellites Designed for Aeronautical Use is Scheduled for Launch in 1999" (1998) 52:9 *ICAO Journal* 16 at 17.

42 WW/IMP-WP/45, *supra* note 40 at 5.

43 P. Hartl and M. Wlaka, "The European Contribution to a Global Navigation Satellite System" (1996) 12:3 *Space Policy* 167 at 171.

44 The role of each organisation in the ETG can be defined as follows: Eurocontrol is responsible for the definition of mission requirements for civil aviation, operational tests, system validation and certification; ESA, for the development and operation of EGNOS; and the European Commission, for institutional and policy matters, including international co-ordination. See European Tripartite Group, "Europe Pursuing a Broad Multimodal Satellite Navigation Programme as its Contribution to GNSS" (1997) 52:9 *ICAO Journal* 13 at 14 [hereinafter ETG].

45 See WW/IMP, "EGNOS Space Based Augmentation Service to GPS and GLONASS", ICAO WW/IMP-WP/67 (11 May 1998) [hereinafter WW/IMP-WP/67] at para.2.1.

46 See *ibid.* For an interesting and recent analysis of the Eurocontrol Convention, see R.D.van Dam, "Recent Developments at the European Organisation for the Safety of Air Navigation (EUROCONTROL)" (1998) XXIII *Annals of Air and Space Law* 311 at 311-320 [hereinafter van Dam].

47 ETG, *supra* note 44 at 13.

48 EU, *Commission Working Document, Sec (1999) 789 final of 7 June 1999, Towards a Coherent European Approach for Space,* [1999], http://europa.eu.int/comm/jrc/space/com_doc_en.html (date accessed: 5 December 1999) at 17 [hereinafter COM Sec (1999) 789].

49 For a more detailed explanation of the GNSS-1 and GNSS-2 systems' concept and mission, see N. Warinsko, "Du GPS au GNSS, Le Point sur la Situation Internationale" (1995) 13 *Le Transpondeur* 19 at 21-24; "Global Satellite Navigation: From GNSS-1 to GNSS-2" (1997) 41 Prospace 2 at 2-5; Y. Trempat, "Les Projets GNSS: La Contribuition Européenne" (1996) 44:173 *Revue Navigation* 41 at 44-51.

50 ETG, *supra* note 44 at 15.

51 See Thomson-CSF, "Egnos: The Future European Navigation System" (1997) 41 Prospace 6 at 6-8; WW/IMP-WP/67, *supra* note 45 at paras.4.4 - 4.8.

52 See *COM (1999) 54, supra* note 32.

53   See COM Sec (1999) 789, *supra* note 48 at 17. "Failure by Europe to act would strengthen the present U.S. market dominance and leave Europe entirely dependent on the U.S. for many security-related matters." UK, Department of the Environment, Transport and the Regions, *Consultation on the European Commission's Communication on Galileo, Involving Europe in a New Generation of Satellite Navigation Services COM (1999) 54 final* (April 1999), http://www.aviation. detr.uk.consult/galileo/index/htm (date accessed: 09 August 1999) [hereinafter UK Consultation].

54   See EU, *Galileo Working Document, Version 1.1 of 7 June 2000, Galileo Definition Phase – Initial Results* [2000], http://www.galileo-pgm.org/indexrd.htm (date accessed: 14 August 2000) at 20 [hereinafter *Galileo*].

55   EU, *Council Resolution of 19 July 1999 on the Involvement of Europe in a New Generation of Satellite Navigation Services – Galileo – Definition Phase,* [1999] O.J.C. 1999/C 221/01.

56   EU, *Communication COM (1998) 29 final of 21 January 1998, Towards a Trans-European Positioning and Navigation Network, Including a European Strategy for Global Navigation Satellite Systems (GNSS)* [1998] Bulletin EU ½ 1998, Transport (1/26) [hereinafter *COM (1998) 29 final*] at 1.3.171.

57   See *COM (1999) 54, supra* note 32.

58   See EU, *Council Resolution of 17 June 1999 on the Commission Communication on "Galileo, Involving Europe in a New Generation of Satellite Navigation Services",* [1999] Bulletin EU 6-1999, Transport (2/9) at 1.2.83.

59   See EU, *Communication COM (2000) 750 final of 22 November 2000, Galileo,* [2000], http//www.galileo-pgm.org/indexrd.htm (date accessed: 5 January 2001) [hereinafter *COM Definition Phase*], Annex 5.

60   Results of the work carried by the Galileo Overall Architecture Definition (GALA) project have shown that albeit the potential architectures in terms of constellation, namely a 30 medium-earth orbit (MEO) constellation and a 24 (MEO) + 8 geostationary satellites, would offer similar global performances, a series of advantages concerning homogeneity, launch strategy, high flexibility in terms of back-ups, simplified replenishment, allocation of orbital slots, lower costs and the ground segment of the MEO-only approach have made it the preferred option of the European Commission. See *Galileo, supra* note 54 at 11.

61   See especially *COM Definition Phase, supra* note 59, Annex 7.

62   1999/C 221/01, *supra* note 55.

63   See *ibid.*

64   See *COM (1998) 29 final, supra* note 56 at 1.3.171.

65   See *Directive of the Russian Federation, supra* note 35; *COM Definition Phase, supra* note 59 at 20.

66   "Galileo might transmit on two of the current GLONASS frequencies and one or more GPS frequencies. Use of frequencies covered by the European filings in the ITU will also be considered". EU, *Commission Communication of 10 February 1999, Galileo, Involving Europe in a New Generation of Satellite Navigation Services, Final Text,* G:\07\02\08\01-EN\final\text.doc [1999] at 12, note 22, http:/www.fma.fi/ radionavigation/doc/galileo2.pdf (date accessed 5 December 1999) [hereinafter *COM Final Text*].

67   See *COM Definition Phase, supra* note 59 at 21; *Galileo, supra* note 54 at 24.

68 See *COM Definition Phase, ibid.* at 11-13; *Galileo, supra* note 54 at 8; Mamlouk, M., "Galileo" (American Bar Association, Forum on Air and Space Law, Montreal, 3 August 2000). See also *GNSSP Report, supra* note 24 at 3.3; *COM Final Text, supra* note 66 at 11, 16, 17.

69 See *COM Definition Phase, ibid.* at 19-20.

70 See *COM Definition Phase, ibid.* at 15.

71 EU, "Get Galileo to Set Pace in Satellite Navigation", Research and Development Sector (10 February 1999), http://www.eubusiness.com/rd/index.htm (date accessed: 5 December 1999) at 2.

72 See *COM Definition Phase, supra* note 59, Annex 7.

73 See *COM Final Text, supra* note 66 at 16, 17. But see UK Consultation, *supra* note 53 at 3,4, which contends that "the possibility that certain uses of Galileo may be made mandatory to generate revenue and to make savings through the withdrawal of conventional aids is of concern. The government considers that user's requirements and benefits need greater investigation and cost-benefit analysis." The private sector is still reticent about participating in the financing of the system, preferring to concentrate on the development of related applications whereby revenues are secure.

74 *COM Final Text, supra* note 66, Annex IV. See especially, N.Warinsko, "Ambitious Project Would Involve Europe in New Generation of Satellite Navigation Services" (1999) 54:9 *ICAO Journal* 4 at 4-5, 29 [hereinafter Warinsko].

75 See *COM Definition Phase, supra* note 59 at 27-28.

76 See *COM Definition Phase, ibid.*

77 See ICAO, *First Meeting of the Secretariat Study Group on Legal Aspects of CNS/ATM Systems*, ICAO SSG-CNS/I-IP/1 (April 1999) at 4.

78 See *COM Final Text, supra* note 66 at 21-22.

79 For an analysis of the impact of Galileo on the satellite navigation market and an estimate of gross economic benefits for Europe, see *ibid.,* Annex IV.

80 See *ibid.* at 26-27.

81 *GNSSP Report, supra* note 24 at para.3.2.1.1.

82 See J.Spiller and T.Tapsell, "Planning of Future Satellite Navigation Systems" (1999) 52:1 *The Journal of Navigation* 47 at 47.

83 See ICAO, *Global Air Navigation Plan for CNS/ATM Systems,* version 1 (Montreal: ICAO, 1998) vol. 1 at para.6.7.1. [hereinafter *Global Plan*].

84 The supplemental-means GNSS must only meet accuracy and integrity requirements for a given operation or phase of flight, as long as it is used in conjunction with a sole-means navigation system on board the aircraft. Requirements for the primary-means GNSS do not differ from those, except that once there is no supporting sole-means navigation system on board, operations must be limited to specific times to ensure safety is not compromised. *ICAO, 156th Session of the Council, Use of GNSS as a Sole Means of Navigation,* ICAO C-WP/11051 (5 February 1999), presented by the Secretary General at 2-3 [hereinafter C-WP/11051].

85 See ICAO, *Third Meeting of the Global Navigation Satellite System Panel,* GNSSP/3 (12-23 April 1999) [hereinafter GNSSP/3], "Use of GNSS as Sole Means of Navigation", ICAO Doc.GNSSP/3-WP/29 (9 April 1999) at 6.7.5.

86 During the World-wide CNS/ATM Systems Implementation Conference, some questions regarding the ability of GNSS to become the sole-means navigation system were raised, and the conference noted that ICAO would consider them in the ongoing

development of SARPs for GNSS. For subsequent discussions within the ICAO Council, see ICAO, *155th Session of the Council, 7th Meeting*, ICAO C-Min 155/7 (22 February 1999) at 7-13.

87 Larsen, *supra* note 3 at 10.

88 See L. Bond, "Global Positioning Sense II: An Update" (1997) 39:4 *Journal of Air Traffic Control* 51 at 53.

89 See *Convention on International Civil Aviation*, 7 December 1944, ICAO Doc.7300/6; UN Doc.15 U.N.T.S.295, art. 89 (entered into force 4 April 1947) [hereinafter *Chicago Convention*].

90 B. Cheng, *General Principles of Law as Applied by the International Courts and Tribunals* (Cambridge: Grotius Publications, 1987) at 31 [hereinafter Cheng]. "The right of a State to adopt the course which it considers best suited to the exigencies of its security and to the maintenance of its integrity, is so essential a right that, in case of doubt, treaty stipulations cannot be interpreted as limiting it, even though these stipulations do not conflict with such interpretation". *The Wimbledon Case*, Dissenting Opinion by Anzilotti and Huber, [1923] PCIJ. Rep. Ser. A. No. 1. at 37.

91 Cheng, *ibid.* at 56. For a related Court decision, see *Carlos Butterfield Case* (U.S. v. Denmark) [1890] 2 Int. Arb. at 1185, 1206.

92 B.D.K Henaku, *The Law on Global Air Navigation by Satellite: A Legal Analysis of the CNS/ATM System* (AST, 1998) at 196 [hereinafter Henaku].

93 See O. Carel, "La Protection des Usagers du GNSS Contre les Interruptions de Service" (1998) 46:182 *Revue Navigation* 213 at 213-218. See also *GNSSP Report, supra* note 24, *Appendix 3A-3 to the Report on Agenda Item 3* at 2.7.1.

94 *Convention for the Suppression of Unlawful Acts Against the Safety of Civil Aviation*, 23 September 1971, ICAO Doc.8966 (entered into force 26 January 1973). Article 1 (d) stipulates that any person commits an offence if he *unlawfully* and *intentionally* destroys or damages air navigation facilities or interferes with their operation, if any such act is likely to endanger the safety of aircraft in flight.

95 *Convention on Offences and Certain Other Acts Committed on Board Aircraft*, 14 September 1963, ICAO Doc.8364 (entered into force 4 December 1969), which applies to acts, whether or not they are offences, that may jeopardise the safety of aircraft or of persons or property therein.

96 See ICAO, *Fourth Meeting of the Secretariat Study Group on Legal Aspects of CNS/ATM Systems*, (April 1999) [hereinafter SSG-CNS/4], "Interference with CNS/ATM Systems – Enforcement in the U.S.", ICAO SSG-CNS/4-WP/2 (8 December 2000); SSG-CNS/4, *ibid.* "Unlawful Interference with CNS/ATM – Australian Practice and Law", ICAO SSG-CNS/4-WP/4 (14 December 2000).

97 See McDonald, *supra* note 1 at 291.

98 See Letter from D. Hinson, FAA Administrator, to A. Kotaite, President of ICAO Council (14 October 1994); Letter from A. Kotaite to D. Hinson (27 October 1994), ICAO State Letter LE 4/4.9.1-94/89, attachment 1 (11 December 1994); Letter from N.P. Tsakh, Minister of Transport of the Russian Federation, to A. Kotaite, President of ICAO Council (4 June 1996), Letter from A. Kotaite to N.P. Tsakh (29 June 1996), ICAO State Letter LE 4/49.1-96/80 (20 September 1996).

99 C-WP/11051, *supra* note 84 at 6.

100 See GNSSP/3, *supra* note 85, WP/29 at para.6.7.6 "A number of aircraft may be approved for sole-means GNSS for particular operations or phases of flight. However,

the air traffic service provider must provide a navigation service to all users as necessary to support all phases of flight. It is therefore necessary to harmonise withdrawal of conventional navaids with the introduction of GNSS navigation service." *Ibid.* "The removal of all conventional air navigation aids [is] an option that should be considered with caution and after consultation with users through the regional air navigation planning process." ICAO, 156th Session of the Council, 11th Meeting, ICAO C-DEC 156/11 (15 March 1999) at 3.

101 See ICAO, *156th Session of the Council, 2309th Report to the Council by the President of the Air Navigation Commission*, ICAO C-WP/11057 (8 March 1999).

3 Frequency Spectrum and Orbital Position Considerations

Introduction

Radio frequency issues deserve special consideration for two significant reasons. Firstly, because the spectrum and the geostationary orbit constitute a limited natural resource which must be used efficiently and economically, they are subject to stringent international telecommunication regulation.[1] Secondly, because of the need to accord the navigation system the highest degree of protection against harmful interference from any other radio source.[2]

Preliminary remarks on the subject matter provide a definition of radio waves as electromagnetic radiation, measured in hertz or cycles, which travels in a straight line at the speed of light and is subject to absorption, diffraction, reflection and diffusion. A distinction is made according to different wave-lengths (frequencies), a group of which is called a band. The radio spectrum is divided into very low, medium, high, very high, ultra-high, super high and extremely high frequencies.[3] A portion of the radio frequency spectrum may be allocated and assigned for a particular role, such as communication or navigation services.[4]

Current technology allows satellite systems to exploit most but not the entirety of the electromagnetic resource across a wide array of frequencies, and still there are particular bands that have especially desirable signal propagation characteristics.[5] Until new technology is developed, difficult allocation choices will have to be made in order to establish eligibility to the use of these scarce orbital resources.[6] The capacity of the spectrum is therefore limited and is liable to become saturated.[7]

Reflecting the dual nature of the orbit/spectrum resource, apart from having a precise orbital position, a satellite must be assigned a specific frequency in order to avoid interference.[8] Harmful interference between satellite transmissions occurs as a result of either physical proximity or the use of the same radio frequency by multiple satellites in the same area.[9] Consequently, legal guarantees must be provided against any such interference by means of agreements at the international level on standards for its operation and co-ordination.[10]

The ITU Regulatory Framework

Access to an orbital position and associated radio frequency, though dependent upon the voluntary action of an individual State to choose and assign a position/frequency to its radio stations, is managed and regulated by the International Telecommunications Union (ITU),[11] a specialised agency of the United Nations. To this end, it has been charged, *inter alia*, with the following functions:

11 (a) [to] effect allocation of bands of the radio-frequency spectrum, the allotment of radio frequencies and registration of radio-frequency assignments and any associated orbital positions in the geostationary-satellite orbit in order to avoid harmful interference between radio stations of different countries;
12 (b) [to] co-ordinate efforts to eliminate harmful interference between radio stations of different countries and to improve the use made of the radio-frequency spectrum and of the geostationary-satellite orbit for radiocommunication services;
...
18 (h) [to] undertake studies, make regulations, adopt resolutions, formulate recommendations and opinions, and collect and publish information concerning telecommunication matters; ...[12]

The basic instrument[13] of the International Telecommunications Union is the Constitution, which is complemented by the Convention of the ITU and Administrative Regulations, namely, Radio Regulations and International Telecommunication Regulations, which all together constitute its regulatory framework.

Whereas the ITU Constitution/Convention establishes its legal structure,[14] stipulates general provisions relating to telecommunication, and creates specific provisions in respect to radio communication,[15] the Radio Regulations[16] address detailed technical issues, procedures and the regulation of the orbit/spectrum usage. Nevertheless, they are all considered to be international treaties and no distinction is made as regards their binding nature.[17] It should be duly noted that unlike the Annexes to the Chicago Convention, ratification, acceptance or approval of the Constitution and the Convention, or accession to these instruments, also constitute consent to be bound by the Administrative Regulations previously adopted at World Radio Conferences.[18]

It is interesting to note that, where appropriate, Annex 10 to the Chicago Convention has actually paraphrased relevant Radio Regulations in its volume II, and specifically emphasised that they should be applied in all pertinent cases,[19] thus indicating the relevance of telecommunication technical legislation directly affecting aviation communication, navigation and surveillance.[20]

The general legal principles with respect to the use of the radio spectrum are contained in the provisions of Article 44 of the ITU Convention which stipulates that:

> [R]adio frequencies and the geostationary-satellite orbit are limited natural resources and ... must be used rationally, *efficiently* and *economically,* in conformity with the provisions of the Radio Regulations, so that countries or groups of countries may have *equitable access* to both, taking into account the *special needs of developing countries* and the geographical situation of particular countries.[21]

Furthermore, States "shall endeavour to limit the number of frequencies and the spectrum used to the minimum essential to provide in a satisfactory manner the necessary services. To that end, they shall endeavour to apply the latest technical advances as soon as possible".[22]

It is the Radiocommunication Sector of the ITU which is tasked with the equitable sharing and efficient use of the radio spectrum and the geostationary satellite orbit, following decisions taken at the World Radio Conferences.[23] The process is called "allocation" and means the entry in the Table of Frequency Allocations of a given frequency band for the purposes of its use by one or more terrestrial or space radiocommunication services.[24] The term should, however, be differed from "assignment" and "allotment" of frequencies, the former meaning an authorisation granted by a national administration for the use of a certain frequency by a radio station, and the latter the entry of a designated frequency channel in an agreed plan, adopted by a competent conference, through which participating member States distribute among themselves the geostationary orbital positions and radio frequencies.[25] Moreover, special agreements on telecommunication matters, which do not concern all members of the Union, can be made by members for themselves or their operating agencies in whatever arena, but shall not be in conflict with the terms of the Constitution, the Convention or the Administrative Regulations.[26]

The ITU regulatory regime governing the sharing of orbit and spectrum resources has been characterised by the development of two procedures applied in different parts of the spectrum, namely, the *a priori* and the *a posteriori* procedures.[27]

Reflecting the "first-come, first-served" principle, the *a posteriori* co-ordination procedure offers a means of achieving the efficient use of orbit/spectrum, not leaving any segment of the resource unused.[28] It seeks to ensure formal recognition and protection against harmful interference from late comers of assigned frequencies and orbital positions by means of a detailed procedure of notification and registration. Such procedure must be concluded so that a new frequency can be entered in the Master Register and be accorded international protection.[29] Nevertheless, even "if an assignment is not in conformity with the Convention and the Radio Regulations, the notifying country may still have it recorded in the Master Register".[30] Thus it has been asserted that the Radiocommunication Sector has no real enforcement power and cannot exercise any control over how member States use their assigned frequencies.[31]

As the demand for the orbit/spectrum resource increased and the perception of its scarcity grew, so did the concerns on the part of the developing world that the technologically advanced States could ultimately monopolise the available frequencies. These concerns would finally be responsible for setting the stage for the development of the present *a priori* allocation system.[32]

An exception to the general rule, the *a priori* planning procedure is based on the need to guarantee the equitable access to the spectrum/orbit resource. By means of frequency/orbit position plans, future rights of use of particular predetermined radio frequencies and orbital positions are granted to all countries, without further need of pre-co-ordination or enquiry about priority issues.

It is important to note that the *a priori* plans adopted so far have been limited to broadcasting and fixed services. In essence, the majority of the orbit/spectrum resource remains accessible on the first-come, first-served basis.[33]

The Outer Space Treaty and the Orbit/Spectrum Resource

Due consideration should be given to the provisions of the Outer Space Treaty,[34] whereby the basic principles of free exploration and use of outer space, and prohibition of claims to sovereignty by individual States are established.[35] In this sense, Article 1 states that "[o]uter space, including the moon and other celestial bodies, shall be free for exploration and use by all States without discrimination of any kind, on a basis of equality and in accordance with international law". Article 2 adds to the concept of the equitable sharing of outer space as a *"res communis"* by providing that it is not subject to national appropriation by claim of sovereignty, by means of use or occupation, or by any other means.

In addition, in its preamble, the treaty refers to the "common interest of all mankind", which concept is undeniably applicable to the equitable use of the radio spectrum and the geostationary orbit, and has actually served as the basis for related discussions in the forum of the ITU.[36]

Accordingly, States share the benefits of the exploitation of outer space on the basis of the principle of equity.[37] As one of the general principles of law recognised by civilised nations,[38] it is part of the law to be applied by the International Court of Justice to decide any disputes submitted to it. It is interesting to note, however, that the most divergent views on the character of such principles have been expressed:

> While some writers regard them merely as a means for assisting the interpretation and application of international treaty and customary law, and others consider them as no more than a subsidiary source of international law, some modern authors look upon "general principles" as the embodiment of the highest principles – the "superconstitution" – of international law.[39]

Directly relevant to the issue of orbit/spectrum resource presently under consideration, the principles above give a clear indication that the recording of an assigned position does not imply national property rights. Therefore appropriation by any State constitutes a direct infringement thereto.[40]

In brief, although no State is allowed to own any orbital position or assigned frequency, all States may use these common resources provided that the international telecommunication regulations and procedures are applied.

A final question may arise as to whether a State which acquires a certain orbital position or radio frequency under an *a priori* plan is allowed to sell such "slots" in exchange for some compensation.[41] It appears a possibility exists as long as the prescribed procedures for modifications in the allotment plans are not contravened.[42]

Concluding Remarks

It is indisputable that all the above-mentioned considerations regarding the radio spectrum and frequency allocation exert direct influence upon the CNS/ATM systems, whose functioning is highly dependent on radio communications and therefore also predicated on the efficient use of radio frequencies. Hence, in order to support the system's safety-of-life applications, guaranteed access to well-protected, interference-free radio frequencies is an absolute prerequisite for the aviation sector.[43]

The basic navigation functions of the GNSS are primarily based on the availability of the 1559 to 1610 MHz Aeronautical Radionavigation Services (ARNS)/Radionavigation Satellite Services (RNSS) frequency band. Accordingly, the GNSS elements (GPS, GLONASS, SBAS and GBAS) all currently operate in this band. As the core frequency for supporting present and future aeronautical applications of GNSS, it deserves absolute protection against harmful interference of any source, especially regarding the potential use of GNSS as a sole means of navigation.[44]

However, the need for conciliatory regulation to satisfy the many divergent requirements for spectrum allocation world-wide has generated a dangerous tendency at the ITU in terms of the reduction of the spectrum available for international civil aviation. Having identified the issue as a problem requiring urgent attention and international co-ordination, ICAO strongly stated its position at the last World Radiocommunication Conference (WRC-2000) on the need for the introduction of an aeronautical safety factor to ensure protection of the 1559-1610 MHz for the exclusive use of the aeronautical radionavigation service and the radionavigation satellite service.[45]

In fact, a most controversial proposal was introduced by INMARSAT at the WRC-1997 to allow mobile satellite services (MSS)(space-to earth) in the 1559-1567 MHz band. No agreement having been reached at the time, a Resolution was passed which called for further technical studies on the compatibility between MSS and RNSS in the referred band. After extensive investigation, the ITU Working Party 8D concluded that sharing between ARNS/RNSS and MSS is not feasible in any portion of the 1559-1567 MHz band. In line with ICAO's position, the WRC-2000 finally agreed that no allocation to the MSS should be made in this band.[46]

Also favourable to civil aviation was the agreement to downgrade the fixed service, which also operates in the GNSS band in a number of countries, to a secondary status by January 2005, when it will be required not to cause any interference to GNSS, which is to have priority over the fixed service.[47]

New allocations to RNSS were also made in various bands, thus enabling the introduction of Galileo and GPS L5.[48] As previously discussed, it is possible that, in the long-term, the GNSS evolution will make use of other new frequency bands, which will also require appropriate protection for adequate GNSS safety-of-life applications.

Nevertheless, the international civil aviation community must bear in mind that a successful long-term outcome shall only be achieved through proper international co-ordination.[49] Continuing availability and guaranteed access to frequency bands for satellite navigation and communication systems, and particularly full protection and exclusive allocation of GNSS frequency bands will require continuous, expedite concerted action by States,[50] as well as agreement at the global level so that the appropriate regulatory provisions are incorporated in the ITU Radio Regulations in that regard.

Notes

1   See B.D.K Henaku, *The Law on Global Air Navigation by Satellite: A Legal Analysis of the CNS/ATM System* (AST, 1998) at 10 [hereinafter Henaku].
2   See W.T.Young, "Potential Interference on the Radio Spectrum Allocated for GNSS Needs Urgent Attention" (1996) 51:7 *ICAO Journal* 25 at 26 [hereinafter Young].
3   N.M.Matte, Aerospace Law: Telecommunications Satellite (Toronto: Butterworths, 1982) at 2.
4   See Henaku, *supra* note 1 at 135.
5   See M.A.Rothblatt, "Satellite Communications and Spectrum Allocation" (1982) 76 *American Journal of International Law* 56 at 56 (LEXIS/NEXIS).
6   See S.A.Levy, "Institutional Perspectives on the Allocation of Space Orbital Resources: The ITU, Common User Satellite Systems and Beyond" (1984) 16 *Case Western Reserve Journal of International Law* 171 at 175 [hereinafter Levy].
7   See R.L.White and H.M.White Jr., The Law and Regulation of International Space Communication (Boston: Artech House, 1988) at 5 [hereinafter White and White].
8   See J.C. Thompson, Comment, "Space for Rent, The International Telecommunications Union, Space Law and Orbit/Spectrum Leasing" (1996) 62 *Journal of Air Law and Commerce* 279 at 280, note 2 [hereinafter Thompson] (LEXIS/NEXIS).
9   See *ibid.* at 284.
10   See R.S.Jakhu, "International Regulation of Satellite Telecommunication" (1991), in *Legal Aspects of Space Commercialisation* (Tokyo: CSP Japan, 1992) [hereinafter Jakhu].
11   For a history of the ITU since its inception in 1865 with the creation of the International Telegraph Union as well as an analysis of its current structure and functioning, see F.Lyall, *Law and Space Telecommunications* (Aldershot: Dartmouth, 1989) at 313-325. See also R.S.Jakhu, "The Evolution of the ITU's Regulatory Regime Governing Radiocommunication Services and the Geostationary Satellite Orbit" (1983) VIII Ann. Air and Sp. L. 381 at 381-406; ITU, "International Telecommunication Union" in

Space Law: Applications, Course Materials (Montreal: McGill University, 1997) at 81ff.

12   *Constitution of the International Telecommunications Union, Geneva, 22 December 1992 (entered into force 1 July 1994), art. 1, Nos. 11, 12 and 18] [hereinafter ITU Constitution].*

13   Throughout the existence of the ITU, its basic document was a result of various incarnations of a single instrument, the International Telecommunication Convention. For example, the Madrid (1932), Nairobi (1982), Nice (1989) and Geneva (1992) conventions. It was the 1989 Nice Plenipotentiary Conference which divided the 1982 Convention into a Constitution and a Convention, the former comprising only provisions of a constitutional nature to which amendments are less likely to be necessary, the latter containing those provisions more likely to change. See F.Lyall, "Communications Regulation: The Role of the International Telecommunication Union" (1997) *Journal of International Law and Technology*, http://elj.warwick.ac.uk/jilt/commsreg/97_3lyal/lyall.TXT (date accessed: 3 December 1999) at para.3.1.

14   The structure of the ITU as defined in the 1992 Constitution shall comprise: "(a) the Plenipotentiary Conference, which is the supreme organ of the Union; (b) the Council, which acts on behalf of the Plenipotentiary Conference; (c) world conferences on international telecommunications; (d) the Radiocommunication Sector, including world and regional radiocommunication conferences, radiocommunication assemblies and the Radio Regulations Board; (e) the Telecommunication Standardisation Sector, including world telecommunication standardisation conferences; (f) the Telecommunication Development Sector, including world and regional telecommunication development conferences; (g) the General Secretariat". *ITU Constitution, ibid.* art. 7.

15   See White and White, *supra* note 7 at 69.

16   Partial or complete revisions of the Radio Regulations are made at the World Radio Conferences (WRC) convened normally every two years. See *ITU Constitution, supra* note 12, art. 13, Nos. 89, 90.

17   See C.Q.Christol, *The Modern International Law of Outer Space* (New York: Pergamon Press, 1982) at 548 [hereinafter Christol]; Henaku, *supra* note 1 at 42-43.

18   See *ITU Constitution, supra* note 12, art. 54, No. 216.

19   See *Convention on International Civil Aviation*, 7 December 1944, ICAO Doc.7300/6; UN Doc.15 U.N.T.S.295,(entered into force 4 April 1947) [hereinafter *Chicago Convention*], Annex 10, vol. II, Introduction.

20   See Henaku, *supra* note 1 at 12, 13, 42.

21   *ITU Constitution, supra* note 12, art. 44, No. 196 [emphasis added].

22   *Ibid.*, No. 195.

23   See *ibid.*, Article 12, No. 78. See also, *Christol, supra* note 17 at 549.

24   See ITU, *Radio Regulations* (1990), No. 17.

25   See *ibid.* Nos. 18, 19.

26   See *ITU Constitution, supra* note 12, art. 42, No. 139. See also, Henaku, *supra* note 1 at 12.

27   For an analysis of the progressive evolution of the regulatory regime and the above-mentioned procedures, see Thompson, *supra* note 8 at 290-302. See also, N.Jasentulyana, "The Role of Developing Countries in the Formulation of Space Law" (1995) XX-II *Annals of Air and Space Law* 95 at 117-122.

28   See Thompson, *ibid.* at 291.

29 "All stations, whatever their purpose, must be established and operated in such a manner as not to cause harmful interference to the radio services or communications of other Members or of recognised operating agencies, or of other duly authorised operating agencies which carry on a radio service, and which operate in accordance with the provisions of the Radio Regulations." *ITU Constitution, supra* note 271, art. 45, No. 197.

30 Jakhu, *supra* note 10 at 117.

31 See Levy, *supra* note 6 at 186.

32 See Thompson, *supra* note 8 at 291.

33 See *ibid.* at 295. See especially R.S.Jakhu, Remarks, "Developments in the International Law of Telecommunications: Strategic Issues for a Global Telecommunication Market" (1989) 83 *American Society of International Law – Proceedings* 385 at 391.

34 *Treaty on the Principles Governing the Activities of States in the Exploration and Use of Outer Space, Including the Moon and Other Celestial Bodies*, 27 January 1967, 610 U.N.T.S. 205 (entered into force 10 October 1967) [hereinafter Outer Space Treaty]. For a detailed analysis of the treaty, see B.C.M.Reijinen, *The United Nations Space Treaties Analysed* (Gif-sur-Yvette Cedex, France: Frontières, 1992), ch. 1 and 3 [hereinafter Reijinen].

35 Evidence of the early acceptance of those principles is provided by the General Assembly Resolution 1962 (XVIII), entitled "Declaration of Legal Principles Governing the Activities of States in the Exploration and Use of Outer Space", adopted unanimously on 13 December 1963.

36 See Reijinen, *supra* note 34 at 9-17. See also E. Chiavarelli, "Satelliti e Sicurezza della Navigazione Aerea: Aspetti Giuridici e Ipotesi di Responsabilità" (1990) XIV:2 *Diritto e Pratica dell'Aviazione Civile* 383 at 386-390.

37 See Reijinen, *ibid.* at 17.

38 *Charter of the United Nations and Statute of the International Court of Justice, 26 June 1945, 16 U.S.T. 1134 (entered into force 24 October 1945),* art. 38, para.1 (c).

39 B. Cheng, *General Principles of Law as Applied by the International Courts and Tribunals* (Cambridge: Grotius Publications, 1987) at 4,5 [hereinafter Cheng]. For further discussion on the subject, see *ibid.* at 1-26; I.Brownlie, *Principles of Public International Law* (Oxford: Clarendon Press, 1998) at 15-19.

40 See Reijinen, *supra* note 34 at 87-102; Jakhu, *supra* note 10 at 120-123. A claim by eight equatorial countries to the sovereignty of the geostationary orbit was laid down in 1976 in the Bogotá Declaration. For a comprehensive discussion on the subject, see S.Gorove, *Developments in Space Law, Issues and Policies* (Dordrecht: Martinus Nijhoff//Kluwer, 1991) at 21-26, 80.

41 For an illustration, see Thompson, *supra* note 8 at 280, where the author describes how the nation of Tonga, after having acquired 16 orbital slots in the early 1990s, proceeded, through its satellite company - Tongsat, and in absolute violation of ITU regulations, to rent an allotment to a Colorado company, and then auctioned off its remaining slots.

42 See J.I.Ezor, "Costs Overhead: Tonga's Claiming of Sixteen Geostationary Orbital Sites and the Implications for U.S. Space Policy" (1993) 24 *Law and Policy in International Business* 915 at 933-935, 941.

43 See Kotaite, A., Address (International Telecommunication Union World Radio Communication Conference 2000, 8 May 2000).

44 See ICAO, *Report on the Third Meeting of the Global Navigation Satellite System Panel*, GNSS/3-WP/66 (12-23 April 1999) at para.3.4 [unpublished][hereinafter *GNSSP Report*].

45 See *ibid.* Appendix D to the Report on Agenda Item 1.

46 See ICAO, *Air Navigation Commission, Report on the Results of the ITU World Radiocommunication Conference (2000)*, ICAO AN-WP/7546 (09 June 2000) [hereinafter AN-WP/7546), Appendix B at 5.

47 See AN-WP/7546, ibid. Appendix B at 1.

48 See *ibid.* at 2. For further details on frequency allocations to ARNS/RNSS, see ITU, *ITU WRC-2000 Provisional Final Acts, 2nd Edition* (9 August 2000).

49 See Young, *supra* note 2 at 26. For the ICAO position for the ITU WRC-2003, see ICAO, *Assembly, 33rd Session, Technical Commission,* "Results of the ITU Radiocommunication Conference 2000 (WRC-2000) and Preparation for Future WRCs", ICAO A33-WP/40 (28 June 2001).

50 For example, comprising 43 European States, the European Conference of Postal and Telecommunications Administrations (CEPT) is responsible for developing and negotiating concerted European positions for the ITU Word Radiocommunication Conferences. See EU, *Commission Communication of 10 February 1999, Galileo, Involving Europe in a New Generation of Satellite Navigation Services, Final Text*, G:\07\02\08\01-EN\final\text.doc [1999], ch. 2 at 22, http:/www.fma.fi/radionavigation/doc/galileo2.pdf (date accessed 5 December 1999).

PART III

LEGAL ASPECTS

4 Introduction

> Each contracting State undertakes, so far as it may find practicable, to: a) Provide, in its territory, airports, radio services, meteorological services and other navigation facilities to facilitate international air navigation, in accordance with the standards and practices recommended or established from time to time pursuant to this Convention.[1]

The language is precise, deliberate. The intention, clear. The emphasis, on the responsibility of States. With these few words, the Chicago Convention, in its Article 28, lay the solid foundation of the international regulation of air navigation.[2] Technological development, however, has ultimately put it to trial. In a different context, it may well demand interpretation and reinterpretation, elucidation of fact and detail to accommodate progress and adapt to the circumstances. But it will most definitely remain the underlying principle of the legal framework which has guided and will continue to guide international air navigation over the next millennium.

In the present chapter, attention shall be devoted to an analysis of the legal implications of Article 28 in the implementation and operation of the CNS/ATM systems, especially regarding the Global Navigation Satellite System. The identification of the relevant legal issues will be supported by a study of the existing legal tools and their legal significance. Consideration will also be given to the elaboration and desirability of a more complex and lasting long-term legal framework for GNSS. Its fundamental principles will be analysed, while emphasis will be given to the issue of liability. As a complement to Chapter 3, reference will also be made to the applicable international law in addition to the Chicago Convention.

Notes

1   *Convention on International Civil Aviation*, 7 December 1944, ICAO Doc.7300/6; UN Doc.15 U.N.T.S.295, art. 28 (entered into force 4 April 1947) [hereinafter *Chicago Convention*].

2   For the origins of regulation of international air navigation in the period preceding the Chicago Convention, see B.D.K Henaku, *The Law on Global Air Navigation by Satellite: A Legal Analysis of the CNS/ATM System* (AST, 1998) at 1-7. For the history and development of air law, see I.H.Ph. Diederiks-Verschoor, *An introduction to Air Law*, 5th rev. ed. (Deventer: Kluewer Law and Taxation, 1993) at 1-12.

5 Existing Legal Tools

The existing corpus of air law is represented by the Chicago Convention and the Annexes to the Convention. Other legal tools include, *inter alia*, the Exchange of Letters, the Charter, the Council Statement of 1994, as well as various recommendations, resolutions, guidelines and guiding principles.

The Chicago Convention

Following the conclusions[1] by the appointed Rapporteur, Dr. Werner Guldimann (Switzerland), in his report to the 28th Session of the Legal Committee in 1992, it has been generally agreed that "there is no legal obstacle to the implementation and achievement of CNS/ATM systems", and that "there is nothing inherent in CNS/ATM systems concept which is inconsistent with the Chicago Convention".[2] Moreover, there is a consensus that "GNSS shall be compatible with the Chicago Convention, its Annexes and other principles of international law".[3]

As the basic constitutional instrument of ICAO and representing a vast codification of public international air law, the Chicago Convention is the primary source of regulation for international civil aviation.[4] Several of its provisions are directly relevant to the present study and will receive a thorough analysis in this chapter as it progresses. Principal among all are the following:

a) Safety of International Civil Aviation (Preamble and Article 44);
b) State Sovereignty (Articles 1 and 2);
c) Airport and Similar Charges (Article 15);

d) Provision of Air Navigation Services and Facilities (Article 28);
e) International Standards and Recommended Practices (Articles 37 and 38); and
f) Financing of Air Navigation Facilities (Chapter XV).

It shall be noted that many principles enshrined in the Convention have been incorporated in other specific legal tools for the CNS/ATM systems, thus constituting essentially restatements or elaborations of the provisions thereof.[5] Such principles will continue to guide the implementation of the CNS/ATM systems and influence the establishment of a long-term legal framework.

International Standards and Recommended Practices – The Annexes

The Law-making Function of the ICAO Council

In a clear indication of its unique quasi-legislative function, according to Article 54 (l), the ICAO Council has the mandatory function to adopt international standards and recommended practices which are designated Annexes to the Convention. Such SARPs shall be concerned with the safety, regularity, and efficiency of air navigation. Particularly relevant to the subject under analysis, they shall deal with, *inter alia*: i) communications systems and navigation aids; ii) characteristics of airports and landing areas; iii) rules of the air and air traffic control practices; and iv) collection and exchange of meteorological information.[6]

The procedure for the adoption of any such Annex or amendment of an Annex requires the vote of two-thirds of the Council and a period of three months after their submission to the contracting States for their entering into force. They shall not become effective, however, if in the meantime a majority registers its disapproval with the Council.[7]

Likewise, under Article 37 of the Convention, contracting States undertake to collaborate in securing the highest practicable degree of uniformity in regulations, standards, procedures in all matters in which such uniformity will facilitate and improve air navigation.

Whereas uniformity is the philosophy underlying the development of technical regulation of international civil aviation, with a view to ensuring safety and security, due consideration must be given to the language of Article 37, whereby it can be inferred that the legal obligation accepted by States is not deemed to be unconditional. Accordingly, no obligation exists for States to comply with SARPs but to the highest practicable degree, and under no circumstances shall they be compelled to adopt and incorporate into their national law any standards or recommendations felt to be, in any respect, at any given time or for any particular reason, impracticable to be complied with.[8]

To this end, the possibility of departure[9] from any international standard or procedure is granted to States by the Convention in its Article 38 which states that:

Any State which finds it impracticable to comply in all respects with any such international standard or procedure, or to bring its own regulations or practices into full accord with any international standard or procedure after amendment of the latter, or which deems it necessary to adopt regulations or practices differing in any particular respect from those established by an international standard, shall give immediate notification to the International Civil Aviation Organisation of the differences between its own practices and that established by the international standard. ... [T]he Council shall make immediate notification to all other states of the difference...

The law is precise. Transparency is the issue at stake. Nonetheless, reality has demonstrated that the effective implementation of SARPs on a global level is a matter of the greatest concern. Very few States notify ICAO of their compliance with or file differences to the standards in the Annexes or their amendments. Therefore, it is not possible to accurately indicate what the state of implementation of any of the 18 adopted Annexes is.[10] In this context, one commentator has gone so far as to blatantly infer that "the detached and cautious approach of ICAO to the issue of enforcement of aviation safety standards may have been motivated by political convenience and reluctance to cause a confrontation with defaulting States", and that "the legal status of ICAO standards would be diluted into a 'desirable guidance material' if there is no authority insisting on compliance in the interests of safety".[11]

In reality, however, ICAO has been dealing with this matter with diligence and caution, provided this is a very delicate issue that demands careful attention and consideration. Recently, progress within the Organisation has provided for an efficient mechanism of safety oversight with mandatory safety audits, which will be discussed below.

In brief, notwithstanding the importance of ICAO's role as the international regulatory authority with respect to civil aviation, a role which should be preserved at all costs, it has been generally agreed that one may but speak of a quasi-legislative function when referring to its powers to enforce the standards in the Annexes. This obligation is therefore left to States through the enactment of appropriate national legislation and the institution of enforcement mechanisms.[12]

Finally, in view of the existence of other regional organisations, equally charged with a regulatory role as regards international air transport and air navigation in their respective jurisdictions, such as the European Union, the Eurocontrol, and the JAA,[13] close co-operation with ICAO is highly desirable so as to provide for the harmonisation of regulations, minimise duplication of efforts, and avoid conflicts of competence in the juxtaposition of roles.[14]

Safety Oversight

Safety Oversight may be defined as a function by which a contracting State ensures the effective implementation of the safety-related SARPs and associated procedures contained in the Annexes to the Convention on International Civil Aviation and related documents.[15]

The first initiative towards safety oversight dates back to 1992 with the introduction by the FAA of the International Aviation Safety Assessment Program.[16] The need for ICAO to adopt the leadership role, in order to avoid a proliferation of individual initiatives[17] and to arrive at a global strategy, prompted the establishment of a programme, operational since 1996, by means of which safety oversight assessments of States were conducted by ICAO on a confidential and voluntary basis. The objective therein was to offer, as necessary, follow-up and technical assistance to States in the implementation of ICAO SARPs in the areas of personnel licensing, operations and airworthiness of aircraft.[18] The findings and recommendations[19] of the assessments were kept strictly confidential as a means of safeguarding the State against adverse economic consequences.[20]

Following the recommendations[21] of the Directors' General of Civil Aviation Conference in 1997, relating to the enhancement of the ICAO safety oversight programme, the 32nd Session of the Assembly adopted a Resolution which:

> Resolves that a universal safety oversight audit programme be established, comprising regular, mandatory, systematic and harmonised safety audits, to be carried out by ICAO; that such universal safety oversight audit programme shall apply to all Contracting States; and that greater transparency and increased disclosure be implemented in the release of audit results.[22]

ICAO Universal Safety Oversight Audit Programme
The Universal Safety Oversight Audit Programme (USOAP), which was brought into effect on 1 January 1999, includes an improved systematic reporting and monitoring mechanism on the implementation of safety-related SARPs which serves to help identify States' deficiencies and recommend the

appropriate remedies.[23] To take a practical example, operational and financial autonomy at the national level as opposed to direct government administration are particularly important in the context of the implementation of the CNS/ATM systems. Accordingly, the establishment of autonomous civil aviation authorities properly empowered to regulate, control and supervise all civil aviation activities in the State has been highly recommended by ICAO.[24]

Many benefits will be gained from further publication and dissemination of the audits' outcome by expanding the information in the summary reports to contain differences to recommended practices, relevant procedures and guidance material. However, the assessed State should have reasonable time to remedy such deficiencies before any information is disclosed.[25] The audits are to be carried out with the consent of the State to be audited, by signing a bilateral Memorandum of Understanding with ICAO, as the principle of sovereignty should be fully respected.[26]

The conduct of mandatory and regular audits can be accommodated within the framework of the Chicago Convention. Incorporated in the mandatory functions of the Council are the duties to:

i) request, collect, examine and publish information relating to the advancement of air navigation ...;
j) report to contracting States any infraction of this Convention;
k) report to the Assembly any infraction of this Convention where a contracting State has failed to take appropriate action within a reasonable time.[27]

Furthermore, the Council has the discretionary power under Article 55 to "conduct research into all aspects of air transport and air navigation which are of international importance, communicate the results of its research to the contracting States", and "investigate, at the request of any contracting State, any situation which may appear to present avoidable obstacles to the development of international air navigation; and after such investigation issue such reports as may appear desirable".[28]

The expansion of the programme to other technical fields, initially to include air traffic services, aerodromes and support facilities and services, was recommended by the DGCA to be considered by the ICAO Council, and will imply wider contacts with different entities and authorities, both private and public.[29]

The assessments carried out to date have confirmed that States are facing serious difficulties in fulfilling their safety oversight obligations. The major deficiencies fall into three categories: i) lack of or inadequate primary aviation legislation and regulations, and enforcement provisions; ii) incomplete or inadequate institutional structure, qualified personnel and financial resources;

and iii) ineffective certification and supervision of commercial air transport operations.[30]

As a rule, any differences from ICAO SARPs identified during the course of the audits, and which still exist when the final reports are issued, are deemed to have been notified to ICAO and are incorporated in the Supplement to the appropriate Annexes.[31]

In response to the principle of sovereignty of States enshrined in the Chicago Convention, it is through the enactment of domestic legislation that States give effect to SARPs.[32] Hence the national Civil Aviation Authority (CAA) must be given proper empowerment to regulate, control and supervise civil aviation activities.[33] In this sense, a question may arise whether, in a given State, there is sufficient legal framework for safety oversight.

First and foremost, any State's primary aviation legislation must contain provisions for the delegation of the necessary authority and the assignment of the corresponding responsibility to the Director of the CAA to develop, revise and issue civil aviation regulations.[34] Secondly, ICAO standards must be properly addressed and incorporated into national law by means of specific aviation regulations.[35] In fact, adequate, clear and unambiguous civil aviation regulations, as well as compliance with ICAO SARPs are deemed essential for a State to achieve international recognition in air transport activities.

Other critical issues which must be covered by the primary legislation are: i) provisions for the enforcement of regulations, namely the definition by law of the penalty to be applied in the event of their infringement or violation; ii) a requirement of all international commercial air transport operations to: a) be conducted under the authority of the State; b) hold an air operator certificate; and iii) have the right of access for inspection to all commercial air transport activities.[36]

It can be well anticipated that the implementation of CNS/ATM will clearly add to the magnitude of the challenge of States to fulfil their responsibility for the promulgation and enforcement of safety regulations in their sovereign territory. States may even encounter difficulties in performing their supervisory functions, in view of the new technology involved, the multinational character of systems' implementation and operation, and lack of sovereign control.[37]

Furthermore, of singular interest to the subject is the assertion that:

> Aviation safety is not produced by governments alone: it is produced collectively by the aviation industry and the government. ... If the government fails in ... the implementation of safety oversight, the aviation industry might regulate and monitor itself... Having a competent and effective regulatory authority therefore is very much in the interest of the aviation industry, apart from its undeniable influence on the level of aviation safety.[38]

The influence of the aviation industry in setting the level of safety can be easily recognised in the way GPS, and in a lesser way GLONASS, are dictating standards for international acceptance and use.[39] A risk exists that the industry will continue to set its own standards, preceding or altogether dispensing with any participation of States under the aegis of ICAO. One commentator has argued that "it would appear that the law-making function of ICAO with respect to the GNSS operational systems will have to follow the practice of the actual signal providers as accepted by the users (i.e. the market) rather than lead in the setting of these standards".[40]

GNSS SARPs

In the development of GNSS SARPs, interoperability to accommodate existing and emerging technology variations has therefore become a major concern in order to guarantee a global, seamless implementation. Moreover, from an economical perspective, it is absolutely necessary to ensure that the different elements are able to work together so that the amount of avionics necessary to support the use of GNSS may be minimised.[41] With a view to ensuring the protection of investment in present navigation systems and allow providers and users to implement changes in a planned and cost-effective manner, a specific protective period of six years of advance notification has been proposed.[42]

A validation process has been established to support the development of SARPs, the main objective of which is to ensure that GNSS SARPs are complete, correct and unambiguous, reflecting known requirements of aeronautical safety, and that practical systems can be developed to satisfy these SARPs.[43] The new approach could serve as a most useful tool to limit the number of differences filed, since standards would better respond to the purposes for which they were created.[44]

Annex 10 is the document that provides SARPs for international aeronautical radio communication and navigation systems. Any amendment to the Annex has to be agreed by States and to follow a very lengthy procedure. Generally speaking, a draft text is prepared by a group of experts, then examined by the Air Navigation Commission, and sent to States for consultation, before being approved by the Council, the whole process taking approximately three years.[45] In order to facilitate this process, it has been decided to develop different levels of SARPs: high level SARPs will be included in Annex 10, whereas detailed technical specifications will be left to technical appendixes, such as ICAO manuals or circulars, the latter not requiring any formal international co-ordination for changes.[46]

Legal Significance

A question which has given rise to much controversy is that of the legal significance of the Annexes to the Chicago Convention. The doctrine is found not to be unanimous as legal opinions widely differ over the legal status of ICAO standards and recommended practices.

For one, Michael Milde points out that, in the very words of the Chicago Convention, SARPs are but "for convenience" designated Annexes to the Convention, thus not constituting an integral part thereof. In addition thereto, "they are not subject to the Vienna Convention on the Law of the Treaties".[47]

This rationale, which has received our support, is further reinforced by a previous assertion on the part of Bin Cheng, when commenting on the issue of the quasi-legislative function of ICAO. In his opinion, in contradiction with the 1919 Paris Convention on the Regulation of Aerial Navigation, the Annexes of which were formulated in completion to that Convention and had therefore identical legal force, the Annexes to the Chicago Convention lack the same legal force as the Convention, and are not binding on States against their will. Accordingly, their application is subject to the conditions stipulated in Articles 37 and 38, whereby States are obliged to comply but to the highest practical degree, or to immediately notify ICAO of any differences between their own practices and that established by the international standard.[48]

Both authors, however, agree on the fact that "international standards are not devoid of legal significance" and that damages for "non-compliance may eliminate the State concerned from any meaningful participation in international air navigation and air transport".[49] In the same vein, Nicholas Mateesco Matte has argued that the standards contained in the Annexes are considered to be "soft law".[50]

These arguments have been opposed on several grounds by many other learned writers. For example, Buerghental goes further to state, *ipsis litteris,* that "since under the Convention, the determination as to what is 'practicable' is for each State to make", "… realistically speaking … [there] is no obligation at all, for a State can always find the necessary 'practical' reasons to justify non-compliance with or deviations from international standards".[51]

Kofi Henaku, on the other hand, advocates that standards adopted in accordance with the Chicago Convention indubitably have independent legal force, hence contracting States are confronted with the obligation of having them enforced,[52] the primary existence of which is confirmed by the prefatory clause of Article 38.[53] Moreover, he argues that, with reference to the observance of a treaty, as a consequence of the principle of *pacta sunt servanda*, "the determination of impracticality to perform must be in good faith"[54]

Finally, a distinction should be made between international standards and recommended practices[55] as regards their legal validity. Whereas the uniform application of a standard has been recognised as *necessary* for the safety and regularity of international air navigation so that Contracting States *will* conform in accordance with the Convention, being compulsory any notification of departure thereof, the uniform application of a recommended practice is simply recognised as *desirable*, and States need but *endeavour* to conform.[56]

Guidelines, Guiding Principles and Other Guidance Material

The Special Committee for the Monitoring and Co-ordination of Development and Transition Planning for the Future Air Navigation Systems (FANS Phase II), in the course of its consideration of acceptable institutional arrangements for the future air navigation systems, developed a set of guidelines with a view to assisting States and regional planning groups to assess the adequacy of the proposed systems.

Subsequently approved by the 28th Session of the Legal Committee, these guidelines were arranged in three sections: i) those of a general nature applying to all CNS systems; ii) specific guidelines relating to AMSS; and iii) specific guidelines relating to GNSS.[57]

A list of guiding principles on institutional and legal aspects of the future air navigation systems was also prepared by the ICAO Secretariat and presented to the Tenth Air Navigation Conference in 1991. A recommendation followed that such guidelines and principles be taken into account in the further study of the institutional and legal aspects of the CNS/ATM systems.[58]

Other guidance material has been produced by the GNSS Panel in the form of "Guidelines for the Introduction and Operational Use of the Global Navigation Satellite Systems"[59] to assist States in reaping benefits from the early implementation of the systems.

It must be duly noted, however, that neither the guidelines nor the guiding principles have legal force *per se*, and thus lack enforceability, their application depending on voluntary compliance. Nevertheless, in the absence of more precise legal rules, they constitute important and timely guidance material, and may provide a basis for the future adoption of binding rules.[60]

Checklist of Items

In the context of the long-term GNSS, the ICAO Legal Committee, at its 29th Session, approved a checklist of items to be considered in contracts for GNSS signal provision with providers of signal-in-space.[61] Of limited normative value, it has been recognised that such items could be further developed in a model contract,[62] where general terms and conditions would be provided, thus ensuring uniformity in case it were to be widely accepted.

On the other hand, a view has been expressed that the absence of a mechanism to impose compulsory clauses would definitely render it difficult to ensure compliance with the model. Furthermore, the primary commercial aspect of GNSS services would make individual parties free to negotiate whatever terms and conditions they saw fit, thus contributing to the complete lack of uniformity, especially by reason of the great number of contracts which would need to be concluded world-wide.[63]

Consequently, it has been asserted that, if ever adopted by the relevant ICAO bodies, a model contract for GNSS could not serve as a substitute for the whole legal framework since it would not address the long-term GNSS in its entirety.[64] However, it might be relevant when it comes to the concept of addressing liability through a chain of contracts between GNSS actors at a regional level.[65]

Statement of ICAO Policy on CNS/ATM Systems Implementation and Operation

On 9 March 1994, the Council of ICAO adopted a policy statement outlining the fundamental precepts to be adhered to in the implementation and operation of the CNS/ATM systems. These are: i) universal accessibility; ii) sovereignty, authority and responsibility of States; iii) responsibility and role of ICAO for the adoption and amendment of SARPs; iv) technical co-operation; v) use of existing organisational structure and institutional arrangements; vi) evolutionary implementation of the GNSS; vii) efficient airspace organisation and utilisation; viii) continuity and quality of services; and ix) reasonable cost allocation to users.[66]

Reflecting the most relevant legal and institutional concerns raised by the international community, the document was derived from the above-mentioned guidelines of the FANS (Phase II) Committee, and represents the general criteria which will certainly serve as the basis for a universally acceptable long-term legal framework.

Although constituting only statements of policy, and therefore not a source of law, and despite their absolute lack of enforceability,[67] these non-binding precepts deserve our careful consideration for their importance in the context of a long-term legal framework, and will be further examined in Chapter 6.

The Exchange of Letters

Introduction

Following a recommendation[68] of the 10th Air Navigation Conference with regard to the development of institutional arrangements as a basis for the continued availability of GNSS, the ICAO Council, at its 134th Session on 11th December 1991, requested the Secretary General, as a matter of urgency, "to initiate, with a view to an early conclusion, an agreement between ICAO and GNSS-provider States, concerning quality and duration of GNSS".[69]

For purposes of this study, preference is given to the expressions "signal-in-space provider States", and "Article 28 States" (or "user States"), in order to draw a clear distinction between States actually providing the GNSS signals, and those providing services based on the use of such signals, as part of their obligation arising out of Article 28 of the Chicago Convention.

At the 29th Session of the Legal Committee, in accordance with a proposal of Dr. Kenneth Rattray (Jamaica), Rapporteur on the item "Consideration, with regard to global navigation satellite systems (GNSS), of the establishment of a legal framework", it was submitted that a transitional arrangement between ICAO and the providers of signal-in-space would enable GPS and GLONASS to be recognised as a component part of the evolutionary approach to the definitive GNSS.[70] Consideration was therefore given to a draft Memorandum of Understanding (MOU) contained in Annex III to the Rapporteur's report as a starting point in the drafting of an international legal instrument. In the Rapporteur's opinion, as will be seen later in this Chapter, such instrument should have the form of an international convention or agreement elaborated under the aegis of ICAO.

The provisions of the draft MOU were related to: i) universal accessibility; ii) duration of services and absence of charges, iii) compliance with ICAO SARPs; iv) responsibility and liability for services; v) provision of information and monitoring by ICAO; and vi) preservation of sovereignty as regards the rights of States to control aircraft operations and enforce safety regulations within their own territory.[71]

According to the Rapporteur, "these provisions would enable adequate assurances to be given to the international community in respect of the legitimate concerns expressed regarding the above matters". Moreover, "this initial start would enable the technology in relation to GNSS to be further developed and for the final form of the system to be crystallised within the legal framework of [an] international convention".[72]

Nevertheless, as expected, it was the provision which placed on the State providing the signal-in-space the responsibility and liability to take all necessary measures to maintain the integrity and reliability of the service and its continuous and uninterrupted performance which came to represent the major obstacle. Views were expressed by both affected States that it constituted "too onerous a burden" and that "the subject was too complex to be dealt with in such a summary fashion",[73] placing serious doubts as to whether it would serve as the desired starting point for negotiations between the parties.

The provision, which was by many considered a fundamental element in the draft agreement, was kept with slight alterations, and the draft text was finally approved by the Committee after some deliberations.[74]

Negotiations continued between ICAO and the provider States. However, it would not be by means of an MOU but through an exchange of letters[75] between the President of the ICAO Council and the FAA Administrator that, in October 1994, the U.S. would finally formalise its offer of the GPS' Standard Positioning Service for use by the international community. Similarly, the offer of the Russian Federation of the provision of a standard accuracy GLONASS channel to the world aviation community would follow suit in a letter from the Minister of Transport dated 4 February 1996, subsequently accepted by the ICAO Council.

Legal Significance

The arrangements with the U.S. and the Russian Federation both satisfy most of ICAO's requirements as expressed in the Council Statement and in the draft agreement. In this regard, as previously stated, services will be made available on a continuous basis, free from direct user charges for a minimum duration of 10 and 15 years, respectively, the U.S. having pledged to give six years' notice of termination of the signals. In addition thereto, the fundamental principle of universal accessibility on a non-discriminatory basis has been incorporated. Due consideration has also been given to the principle of sovereignty of States as both letters expressed not to be the intention to limit the rights of any State to control the operations of aircraft and enforce safety regulations within its sovereign territory. Both States have pledged full co-operation with ICAO in the development of SARPs and expressed their expectancy that these would be made compatible with their respective systems. Again, it has been made clear that States will be left free to implement augmentation systems if desired. Furthermore, both undertook to provide ICAO with operational information on their respective systems.[76]

It should be duly noted, however, that neither offer has addressed the complex issue of liability, having limited to state, with similar language, that all necessary measures will be taken to maintain the integrity and the reliability of the services provided. In need of elucidation, the matter has been subject to countless, intense debates.

One commentator has ventured to compare the apparent ambiguity surrounding the U.S. position to an actual disclaimer of liability which, in his view, would be recognised as valid by international air law. He argues that "in the same manner that private legal persons are accorded party autonomy in their contractual relations, the equality of States is recognised as a basic principle determining the character of inter-State relations".[77] In this sense, he goes further to invoke the provisions of the Vienna Convention on the Law of the Treaties in respect to the freedom of States to enter into any agreement and make reservations thereof,[78] as well as the possibility to employ various other exclusionary mechanisms to exclude or limit liability. Although bearing in mind that, in accordance with the principle *pacta tertii nec nocent nec prosunt* enshrined in Article 34 of the Vienna Convention, a treaty cannot create rights or obligations to a third party without its consent, the author infers that users (aircraft operator or passenger) would be in a quasi-contractual relation with the provider of signal-in-space. Therefore, it could be presumed that they had knowledge of the disclaimer and hence would be bound by it.[79]

The issue appears to have been clarified by the U.S. representative on the ICAO Council, who has claimed that the wording "does not mean that the provider may not be held liable for negligent failure of the system".[80] In the same vein, ICAO's Legal Bureau has manifested its opinion in the sense that should an accident occur, "the relevant rules of liability will apply and the signal providers will be held responsible through recourse to the laws of the relevant State".[81]

Notwithstanding the above elucidation and legal rationale, scepticism prevails and many States feel there is still some cause for concern. The issue of liability therefore deserves further analysis and will be studied in Chapter 7.

The legal significance of the Exchange of Letters has led to a variety of legal opinions. Even before the first offer was ever formalised, a view had already been put forward at the Legal Committee that, regarding a transitional arrangement, "the title of th[e] instrument was largely immaterial and that a memorandum of understanding or an exchange of letters would have the same legally binding force among the parties, and that it was the content of the instrument which was of paramount importance".[82]

Opinions were also expressed in the sense that ICAO lacks the powers to enter into legally binding undertakings on behalf of the international civil aviation community.[83] This understanding is shared, for example, by one writer who invokes Article 65 of the Chicago Convention to prove that the ICAO Council has no legal authority to enter into a formal agreement concerning the GNSS.[84] According to said provision, the Council may enter into agreements with other international bodies for the maintenance of common services and for common arrangements concerning personnel. In addition thereto, with the approval of the Assembly, it may enter into any such other arrangements as may facilitate the work of the Organisation.[85] In this author's view, "it would appear impermissible to extend its applicability to the provision of the GNSS". Furthermore, he states that Chapter XV of the Chicago Convention also does not give any such authority to the Council. Consequently, "these unilateral statements and exchanges of correspondence with ICAO do not represent an international agreement".[86]

A view to the contrary has been expressed by another commentator who, citing Schermers and Blokker,[87] submits that "international organisations have competence to enter into international agreements", a fact which has been confirmed "in practice and in judicial decisions. ... [Such] agreements are binding on them, and depending on the nature of the agreement, on the member States".[88]

Again, it has been inferred that an exchange of letters constitutes a promise or unilateral act. As such, they require no *quid pro quo* and might be capable of creating legal obligations, being enough that "a State willingly undertakes to engage in a specified conduct".[89] According to one writer, "a promise or declaration or any sort of international commitment made by a State may be presumed to be a genuine commitment".[90] However, argues another author, "a great deal will depend on the context in which a promise or protest occurs, including the surrounding circumstances".[91] Therefore, "the detection of an intention to be legally bound, and of the structure of such intention, involves careful appreciation of the facts".[92]

Other commentators have stated that "letters exchanged through diplomatic channels are not intended to be legally binding, and are not considered treaties because they do not describe legal obligations in detail".[93] In their view, they would most likely be characterised as a non-binding international agreement, hence not enforceable in law.[94] Moreover, "rules concerning compliance, modifications and withdrawal from treaties do not apply. Nevertheless, these agreements may be considered morally and politically binding by the parties, and the President may be making a type of national commitment when he enters one".[95]

In addition thereto, it has been argued that "if intended to be legally binding, proper U.S. procedures for entering into executive agreements would have to be followed".[96] There is, however, a clear distinction between executive agreements and unilateral policy statements, since only the former "are to all intents and purposes binding treaties under international law".[97]

In terms of result, these opposite views appear to converge to a consensus when considering the actual wording of the U.S. Letter, whereby it rests manifest that it was purportedly submitted "in lieu of agreement", and therefore there was no intention on the part of the American government to conclude a formal international agreement.[98] Furthermore, both letters make reference to constituting but a "mutual understanding"[99] between the parties.

In brief, since it is the common intention of the parties and the spirit, rather than the literal meaning of a treaty which have to be observed,[100] it is clear that the Exchange of Letters has no legal binding effect between the States providing signal-in-space and ICAO, nor in relation to its member-States as third parties. It is therefore submitted that the international community will have to rely on the principle of "good faith",[101] as a safeguard against the availability, continuity, integrity and reliability of the signals provided. Indeed, it may be said that, at least at present, the very success of the implementation of the CNS/ATM systems is largely dependent upon the degree of good faith with which such promises are kept so that confidence placed upon them might prevail in the relations between providers and users.

Charter on the Rights and Obligations of States Relating to GNSS Services

Introduction

On 6 December 1995, pursuant to a request of the 31st Session of the ICAO Assembly, in its Resolution A31-7, the Council established the Panel of Experts on the Establishment of a Legal Framework with Regard to Global Navigation Satellite Systems (LTEP). Within its terms of reference was the mandate "to consider different types and forms of the long-term legal framework" and "to elaborate the legal framework which would respond, inter alia, to the fundamental principles set out in paragraph 6 of the Rapporteur's Report"[102] to the 29th Session of the Legal Committee.

As a result of the discussions during its first meeting in November 1996, and taking into account the recommendation of the Legal Committee that a two-stage approach be followed in the implementation of GNSS, namely the development of a legal framework to permit the implementation of the existing system and the elaboration of a more complete and lasting instrument for the future,[103] the Panel established two working groups to assist in the preparation of the relevant documents and principles.

Accordingly, a Working Group on GNSS Principles (Working Group I), under the Chairmanship of Dr. Kenneth Rattray, was mandated with developing draft provisions of a Charter formulating the fundamental principles for GNSS.[104] A second Working Group was tasked with formulating draft legal principles and provisions on specified legal issues.[105]

The Charter, whose text was approved by the LTEP at its second meeting in November 1997, embodies certain fundamental principles to be observed in the implementation and operation of GNSS. These include: i) the safety of international civil aviation; ii) universal accessibility of GNSS without discrimination; iii) preservation of States' sovereign rights; iv) continuity, integrity, availability and reliability of services; v) international co-operation, among others.[106] Again, no reference is made to the issue of liability for GNSS, since no agreement could be reached on the appropriateness of including a related statement therein.[107]

The principles contained in the Charter do not differ in substance from those previously developed and embodied, in whole or in part, in other documents, such as the Chicago Convention, the Council Statement, the Exchange of Letters, the Guidelines of the FANS (Phase II) Committee, as well as outer space conventions and declarations.[108] A product of consensus, they may be as well an indication of the basic principles which will form part of the long-term legal framework for future GNSS.

A long discussion ensued on which course of action would be recommended with regard to the form of the Charter, which could be given effect in either an international convention or an Assembly resolution. In this regard, a number of experts in the Panel believed that since the Charter was considered as the restatement of existing principles contained in the Chicago Convention, it was not necessary to have another convention to restate these principles, and therefore sustained it should take the form of an Assembly resolution.[109]

On the other hand, a large group of experts did not accept that "the delay to be incurred in the adoption and ratification of a convention should be considered a valid reason for not having such a convention". They were of the opinion that "from a strictly legal point of view, only an international convention could give the principles of the Charter the required binding force" and "maintain the integrity of the legal framework for GNSS".[110] However, in view of lack of consensus, and taking into account that both forms were not mutually exclusive, they had no difficulty in accepting an Assembly resolution as an interim solution or transitional arrangement. Meanwhile, work towards an internationally binding instrument would proceed.[111]

During its 153rd Session, in March 1998, the Council decided to have the Draft Charter submitted to the 32nd Session of the Assembly for adoption.[112] The text was presented next at the World-wide CNS/ATM Systems Implementation Conference in Rio de Janeiro, where a conclusion was reached that "the adoption of the Charter should ... be regarded only as an interim framework for the short-term, while further consideration is given to the long-term legal framework, including consideration of the development of a draft international convention for this purpose".[113] At the next session of the ICAO Assembly in September, 1998, the Charter was therefore framed by the Legal Commission in the form of a resolution and subsequently adopted by consensus by the Assembly.[114]

Legal Significance

Adopted in the form of an Assembly resolution, the Charter cannot be accorded any legal force and therefore must be regarded as legally not binding. Some commentators, having expressed serious doubts as to the usefulness of the instrument, seem to be somewhat displeased with the nomenclature employed which would be indicative of a legal instrument of fundamental importance.[115]

Some views to the contrary have also been expressed that the Charter may constitute obligatory norms of international law, as evidenced by some Assembly resolutions adopted in the past, considered as the statement of customary rules, independently of any treaty.[116]

On the other hand, it could be said that the Charter finds its "legitimacy" in the strong political weight carried by a resolution of the ICAO Assembly as well as in its high persuasive value.[117] Despite its lack of enforceability, it has its merit for reaffirming legal principles of fundamental importance which may constitute the basis for a future binding instrument, and even lead the way towards the adoption of an international convention.

In such a particular context, where legal aspects find themselves intrinsically associated with intricate policy considerations, political and economic affairs of States, the adoption of the Charter as an interim solution reflects the pressing need to create confidence in GNSS without delaying the implementation of the system. Hence, it represented a necessary political step in the interest of the international community still in search of international safeguards to a system not under its control.

LTEP Recommendations

The Working Group on GNSS Framework Provisions (Working Group II) was established by the LTEP with the following mandate:

  a) to analyse and, as appropriate, to draft legal principles or where possible provisions, on the following matters:
    i)   certification;
    ii) liability, including the allocation of liability among the participants in the system;
    iii) administration, financing and cost recovery; and
    iv) future operating structures.[118]

The Group, chaired by Dr. Emilia Chiavarelli (Italy), held its first meeting in April 1997, when it agreed on several legal principles concerning the issues in the terms of reference as a basis for further study.[119] These principles, along with the results of an informal survey[120] conducted through a questionnaire, and additional working papers submitted by the experts were taken into account in the development of a set of recommendations[121] drafted and approved by the Group at its second and third meetings, in September 1997 and February 1998.[122] With the exception of recommendation 11 bis on liability which was adopted by a majority, all recommendations were adopted by consensus.[123]

In the course of its meetings, the LTEP also considered the substance of these recommendations which were, after a few minor amendments, adopted by consensus by the Panel.[124]

Recommendations 1 to 8 are concerned with issues of certification, whereas recommendations 9 to 11 with the issue of liability. Recommendations 12 to 14 deal with administration, financing and cost recovery, and recommendations 15 and 16 with future operating structures.[125]

Despite the vastly divergent viewpoints expressed in the course of the meetings, reflecting the different perspectives and concerns of provider and user States, these recommendations represent a major achievement as regards the necessary first stage of non-binding norms in the long law-making process of any future legal instrument for the long-term GNSS.

In this sense, the President of the ICAO Council had appealed to the panel to work in a spirit of co-operation and compromise in order to find pragmatic solutions for those legal issues. Solutions which, in his own words, "should not impose undue obligations upon the provider States of GNSS services, [but] should nevertheless offer appropriate safeguards for user States".[126]

The World-wide CNS/ATM Systems Implementation Conference

Conclusions and Recommendations

The World-wide CNS/ATM Systems Implementation Conference was convened by ICAO in Rio de Janeiro, Brazil, from 11 to 15 May, 1998. As an action-oriented meeting, its unique aspect consisted in bringing together all major partners in civil aviation, from top-level government, industry decision makers and directors of civil aviation authorities to heads of financial institutions and investors, major manufactures, service providers and users, to consider two critical issues: the financial aspects and the institutional framework for CNS/ATM systems.[127]

As far as financing is concerned, the primary objective of the Conference was to convince service providers and financial institutions that implementation of the systems would generate a significant positive return on investment, which could be recovered through user charges, and that it could be of benefit to lenders, borrowers and users alike.[128] In this respect, the special economic and financial circumstances in many a region on the planet, where a significant majority of States requires assistance, gain particular relevance in the context of the implementation of a seamless, globally co-ordinated and fully interoperable CNS/ATM system.[129]

Recognising that the organisational structure under which CNS/ATM systems are to operate is fundamental to their financial viability, the Conference considered various options at the national and multinational levels, recommended the establishment of autonomous authorities and acknowledged the need to adopt a co-operative, multinational approach to implementing regional and global elements of the systems.[130]

Being not the aim in this chapter to delve too much into the Conference's deliberations, but merely to illustrate how, and to what extent its results will influence decisions on immediate concerns and guide further work on the development of the long-term legal framework, suffice it to say that the Conference arrived at significant conclusions and agreed on recommendations concerning substantive financial, institutional, legal and technical-co-operation aspects of the systems, as well as training needs. Particularly, it recommended that:

> The complex legal aspects of the implementation of CNS/ATM systems, including GNSS, require further work by ICAO. Such further work should seek to elaborate an appropriate legal framework to govern the operation and availability of CNS/ATM, including the consideration of an international convention for this purpose. Such further work should not, however, delay implementation of CNS/ATM systems.[131]

In addition, in carrying it out, the main objective should be to develop and build mutual confidence among States regarding CNS/ATM systems.[132]

Having endorsed the central role of ICAO through the development of technical and operational SARPs, the Conference concluded that "regional arrangements may contribute to the development of a global legal and institutional framework with regard to long-term GNSS, provided they are compatible with the global framework and support the interoperability of regional CNS/ATM components".[133]

Indeed, at the 32nd Session of the ICAO Assembly, a resolution was adopted instructing the Council and the Secretary General, within their respective competencies, and beginning with a Secretariat Study Group, to:

a) ensure the expeditious follow-up of the recommendations of the world-wide CNS/ATM Systems Implementation Conference, as well as those formulated by the LTEP, especially those concerning institutional issues and questions of liability; and

b) consider the elaboration of an appropriate long-term legal framework to govern the operation of GNSS systems, including consideration of an international Convention for this purpose, and to present proposals for such a framework in time for their consideration by the next ordinary Session of the Assembly.[134]

Declaration on Global Air Navigation Systems for the Twenty-first Century

Adopted at the closing of the Conference, and consolidating its conclusions and recommendations, the "Declaration on Global Air Navigation Systems for the Twenty-First Century", of mere informative value, purports to give the world community knowledge about the results of the Conference's work by declaring, inter alia, that:

a) increasing levels of co-operation at the national, sub-regional and global levels will be necessary to ensure transparency and interoperability between systems' elements;

b) the operation of air navigation services by autonomous authorities may contribute to significant economies, increased efficiency and transparency;

c) financing and operation of CNS/ATM systems can be of common benefit to lenders, borrowers and users;

d) sound financial management is critical to securing financing for CNS/ATM projects;

e) planning and implementation of the systems should be on the basis of homogenous air traffic management areas and major international traffic flows, taking into account the diversity of technology.[135]

The Declaration also directly supported the conclusions and recommendations on the legal aspects of CNS/ATM systems, as well as the adoption of the Charter as an interim framework for the short-term, while consideration is given to the long-term legal framework.

Notes

1 See ICAO, *Legal Committee, 28th Session*, Rapporteur's Report on "The Institutional and Legal Aspects of the Future Air Navigation Systems", by Werner Guldimann, ICAO LC/28-WP/3-1 (24 January 1992) [hereinafter *Guldimann Report*] at para.7.1.

2 ICAO, *Report of the 28th Session of the ICAO Legal Committee*, ICAO Doc.9588 – LC/188 (1992) at para.3-12 [hereinafter *Report of the 28th Session*].

3 ICAO, *Global Air Navigation Plan for CNS/ATM Systems*, version 1 (Montreal: ICAO, 1998), vol.1 at para.11.1.1.1 [hereinafter *Global Plan*].

4 See M.Milde, "The International Flight Against Terrorism in the Air" (Tokyo Conference, 3 June 1993) [unpublished].

5 *Global Plan, supra* note 3 at para.11.2.1.

6 See *Convention on International Civil Aviation*, 7 December 1944, ICAO Doc.7300/6; UN Doc.15 U.N.T.S.295, art. 37 (entered into force 4 April 1947) [hereinafter *Chicago Convention*].

7 See *Chicago Convention, ibid.*, art. 90.

8 For example, "due to the lack of funds, personnel, equipment, etc., within the time specified for the application of a new standard". M. Milde, "Aviation Safety Standards and Problems of Safety Audits" (Soochow University Seminar, Taipei, 28 June 1997) [unpublished] [hereinafter Milde].

9   See especially, ICAO, Council, 11th Session, Proceedings of the Council, Part II (1950), *Principles Governing the Reporting of Differences from ICAO Standards, Practices and Procedures*, ICAO Doc.7188 – C/828.

10 ICAO, *Assembly, 31st Session*, "Implementation of ICAO Standards and Recommended Practices", ICAO Doc.A-31 WP/56 (1 August 1995).

11 Milde, *supra* note 8.

12 See M.Milde, "The Chicago Convention – Are Major Amendments Necessary or Desirable 50 Years Later" (1994) XXI:I *Annals of Air and Space Law* 401 at 425.

13 Representing the regulatory authorities of a number of European Civil Aviation Administrations, the Joint Aviation Authorities (JAA) function as an associated body of the European Civil Aviation Conference with the purpose of co-operating in the development and implementation of common aviation safety standards and procedures. See Groenewege, A., *Compendium of International Civil Aviation*, 2nd ed. (Montreal: IADC, 1998) at 229.

14 See F.P. Schubert, "Organisations Régionales et Gestion de la Circulation Aérienne: Réflexion Critique sur le Régionalisme Européen" (1995) XX:I *Annals of Air and Space Law* 377 at 380.

15 See ICAO, *Safety Oversight Assessment Handbook*, 4th ed., 1997 at 1.2 [hereinafter *Safety Oversight Assessment Handbook*].

16 See ICAO, *Directors General of Civil Aviation Conference on a Global Strategy for Safety Oversight* [hereinafter DGCA], "Relationship of the U.S. Federal Aviation Administration's International Aviation Safety Assessment Program to ICAO's Safety Oversight Program", ICAO DGCA/97-IP/1 (3 November 1997), presented by the U.S. "The purpose of the IASA program is to ensure that all foreign carriers that operate to and from the U.S. are licensed under conditions meeting ICAO SARPs and receive adequate continuing safety oversight from a competent CAA". *Ibid.* at para.2.3. See also, ICAO, *6th Meeting of Directors of Civil Aviation - ICAO South American Region*, RAAC/6-IP/4. The FAA classifies the status of a country after its assessment into three categories, namely, i) category I, for those who comply with ICAO standards, ii) category II, for partial compliance, when corrective measures are being implemented; and iii) category III, for those with unacceptable ratings of compliance.

17 For information on other international initiatives, see DGCA, *ibid.*, ICAO DGCA/97-WP-1 at para.7.1ff.

18 See ICAO, *Assembly, 32nd Session, Executive Committee*, "Transition to the ICAO Universal Safety Oversight Audit Programme", ICAO Doc.A-32-WP/61 (6 July 1998) para.2.1 [hereinafter A-32-WP/61].

19 Following an assessment, as agreed upon between ICAO and the assessed State through a Memorandum of Understanding (MOU), a confidential interim report containing ICAO's findings and recommendations was made available to the assessed State, which then undertook to submit to ICAO an action plan addressing such deficiencies. Upon its receipt, a confidential final report was produced by ICAO in response to the action plan, outlining any outstanding differences to ICAO SARPs in Annexes 1, 6, and 8. A non-confidential summary report providing general information could be made available to other States upon request. See DGCA, *ibid.*, "Safety Oversight Today", ICAO DGCA/97-WP 1 (1 October 1997) at para.6.9. On confidentiality issues, see especially DGCA, *ibid.*, "Dealing with Confidentiality Issues", ICAO DGCA/97-WP-4 (2 October 1997). For an example of an MOU, see DGCA, *ibid.*, Appendix.

20 See DGCA, *ibid.*, "The ICAO Safety Oversight Programme, A Quality Assurance Approach to Safety", ICAO DGCA/97-IP/6 (23 October 1997) at para.2.4.

21 See DGCA, Conclusions and Recommendations, ICAO DGCA/97-CR 1 to 8 [hereinafter DGCA/97-CR].

22 ICAO, *Assembly, 32nd Session*, CD-ROM (Montreal, 1998), *Establishment of an ICAO Universal Safety Oversight Audit Programme*, Res. A32-11 at para.1[hereinafter *Res. A32-11*].

23 See *Res. A32-11, ibid.* at paras.2, 3. States may request assistance from ICAO to develop action plans to rectify deficiencies.

24 See DGCA/97-CR, *supra* note 21, ICAO DGCA/97-CR/7 at para.2.1 (a). See especially, ICAO, *Assembly, 32nd Session, Executive Committee*, "Report on Financial and Organisational Aspects of the Provision of Air Navigation Services", ICAO Doc.A-32-WP/49, EX/18 (3 July 1998) at para.4.1. For a comprehensive review of the topic, see below, Administration at 140.

25 See DGCA/97-CR, *ibid.*, ICAO DGCA/97-CRs/4 and 5 at para.2.1 (d).

26 See *Res. A32-11, supra* note 22 at para.3. See also, A-32-WP/61, *supra* note 18 at paras.4.2.3, 4.2.4.

27 *Chicago Convention, supra* note 6, art. 54 (i), (j), and (k).

28 *Chicago Convention, ibid.*, art. 55 (c), (e).

29 See DGCA/97-CR, *supra* note 21, ICAO DGCA/97-CR/6 at para.2.1 (a). See also, DGCA, *supra* note 16, "Expansion of the ICAO Safety Oversight Programme to Other Technical Fields", ICAO DGCA/97-WP-6 (3 October 1997).

30 See A-32-WP/61, *supra* note 18 at para.2.3. See especially DGCA, *supra* note 21, "Results from the ICAO Safety Oversight Program", ICAO DGCA/97-WP-2 (1 October 1997).

31 See *ibid.* at para.2.4.

32 ICAO, *Panel of Experts on the Establishment of a Legal Framework with regard to GNSS, Working Group on GNSS Framework Provisions (Working Group II)*, LTEP-WG/II (22-25 April 1997) [hereinafter LTEP-WG/II], "Legal Aspects of GNSS Certification and Liability", LTEP-WG/II-WP/8 (18 April 1997), presented by O.Carel, M,Denney, E.Hoffstee, P.O'Neill, T.Nordeng., W. t'Hoen, A.Watt, G.White [hereinafter LTEP-WG/II-WP/8].

33 See A.Quiroz, "ICAO Safety Oversight Programme – An Overview" (Senior Civil Aviation Management Course, Lecture, International Aviation Management Training Institute, 8 June 1999).

34 See *Safety Oversight Assessment Handbook, supra* note 15 at para.3.1.1.

35 For example, the FAA's Federal Air Regulations (FARs) together with the British Civil Aviation Regulations form the basic structure of the new Joint Air Regulations (JARs) enacted by the JAA, which are also found to be incorporated into national aviation codes of a number of other third countries. See S. Mattews, "European Air Safety in the New Millennium", in World Market Series, *Business Briefing: European Civil Aviation and Airport Development* (World Markets Research Centre, 1999) 105 at 108 [hereinafter WMRC].

36 See WMRC, *ibid.* at para.3 ff.

37 See LTEP-WG/II, *supra* note 32, "Legal Aspects of GNSS Certification", LTEP-WG/II-WP/2 (18 March 1997) at para.7.

38 DGCA, *supra* note 21, "Safety Oversight, an International Responsibility", ICAO DGCA/97- IP/5 (20 October 1997) at para.8.2, presented by the Kingdom of the Netherlands.

39 Both the U.S. and the Russian Federation have expressed in the letters exchanged with ICAO their willingness that ICAO SARPs be developed to be compatible to their respective systems. See Letter from D. Hinson, FAA Administrator, to A. Kotaite, President of ICAO Council (14 October 1994); Letter from A. Kotaite to D. Hinson (27 October 1994), ICAO State Letter LE 4/4.9.1-94/89, attachment 1 (11 December 1994); Letter from N.P. Tsakh, Minister of Transport of the Russian Federation, to A. Kotaite, President of ICAO Council (4 June 1996), Letter from A. Kotaite to N.P. Tsakh (29 June 1996), ICAO State Letter LE 4/49.1-96/80 (20 September 1996) [hereinafter Letters].

40 M.Milde. "Solutions in Search of a Problem? Legal Aspects of the GNSS" (1997) XXII:II *Annals of Air and Space Law* 195 at 203 [hereinafter Milde]. For an example as regards the introduction of the "FNS-1 package" or the "FANS-A package", see O. Carel and J.L.Jonquière, "Les Spécifications des Systèmes Complexes et Leur Validation" (1999) 47:185 *Revue Navigation* 12 at 19 [hereinafter Carel and Jonquière].

41 See ICAO, *Report on the Third Meeting of the Global Navigation Satellite System Panel*, GNSS/3-WP/66 (12-23 April 1999), Report on Agenda Item 1 at para. 1.1.3 [unpublished][hereinafter *GNSSP Report*].

42 See *ibid.*, Report on Agenda Item 3 at para.4.1.1.

43 See *ibid.*, Report on Agenda Item 1 at paras.1.5.1ff. The methodology includes inspection, testing, simulation and/or analysis.

44 See ICAO, *Report of the First Meeting of the Working Group on GNSS Framework Provisions (Working Group II) of the Panel of Legal and Technical Experts on the Establishment of a Legal Framework With Regard to GNSS (LTEP)*, ICAO LTEP/2-WP/3 (15 September 1997) at para 1:13 [unpublished] [hereinafter *WG/ II Report*].

45 See Carel and Jonquière, *supra* note 40 at 18.

46 See *ibid.* See also G.V.Kinal and F.Ryan, "Satellite-based Augmentation Systems: The Need for International Standards" (1999) 52:1 J.Navigation 70 at 71; *GNSSP Report*, *supra* note 41, Report on Agenda Item 1 at 1.1.4. See especially, ICAO, *Assembly, 32nd Session, Consolidated Statement of ICAO Continuing Policies and Associated Practices Related Specifically to Air Navigation*, Res. A32-14 [hereinafter *Res. A32-14*], Appendix A at para.4, which provides that "SARPs and PANS shall be drafted in clear, simple and concise language. Furthermore, for complex systems, SARPs shall, to the extent possible, consist mainly of broad, mature and stable provisions. For such systems, detailed technical requirements and specifications shall be appendixes to Annexes or be placed in separate documents."

47 Milde, *supra* note 8 at 4-6.

48 See B.Cheng, *The Law of International Air Transport* (London: Stevens, 1962) at 64 [hereinafter Cheng].

49 Milde, *supra* note 8 at 5.

50 See N.M.Matte, "The Chicago Convention, Where From and Where To, ICAO?" (1994) XXI:I *Annals of Air and Space Law* 371 at 378.

51 T.Buergenthal, *Law-Making in the International Civil Aviation Organisation* (Syracuse, New York: Syracuse University Press, 1969) at 78.

52   See B.D.K Henaku, *The Law on Global Air Navigation by Satellite: A Legal Analysis of the CNS/ATM System* (AST, 1998) at 36 [hereinafter Henaku].

53   See *ibid.* at 56-63.

54   *Ibid.* at 55.

55   Mention should also be made of the Procedures for Air Navigation Services (PANS), which mainly comprise procedures intended for world-wide application but regarded as not yet having attained a sufficient degree of maturity for adoption as SARPs, as well as material considered to detailed for incorporation in an Annex. Regional Supplementary Procedures (SUPPS) are, in turn, intended only for application in specific regions. See Cheng, *supra* note 48 at 70-71.

56   See *Res. A32-14, supra* note 46, Appendix A.

57   See *Report of the 28th Session, supra* note 2 at para.3-12. For specific comments on the guidelines, see ICAO, *Legal Committee, 28th Session,* "General Information and Comments Resulting From FANS (II)/3", ICAO LC/28-WP/3-5 (7 May 1992).

58   See ICAO, *Report of the Tenth Air Navigation Conference,* ICAO Doc.9583 - AN-CONF/10 (5-20 September 1991), Recommendation 4/1 at para.4.4.5. [hereinafter *AN-CONF/10 Report*].

59   See ICAO, *Guidelines for the Introduction and Operational Use of the Global Navigation Satellite System,* ICAO Circ.267 [hereinafter *GNSS Guidelines*].

60   See ICAO, *Panel of Experts on the Establishment of a Legal Framework With Regard to GNSS,* LTEP/1 (25-30 November 1996) [hereinafter LTEP/1], "Different Types and Forms of the Long-Term Legal Framework For GNSS", LTEP/1-WP/5 (20 September 1996)[hereinafter LTEP/1-WP/5]. See also A.Kotaite, ICAO's Role with Respect to the Institutional Arrangements and Legal Framework of Global Navigation Satellite System (GNSS) Planning and Implementation (1996) XXI:II *Annals of Air and Space Law* 195 at 198 [hereinafter Kotaite].

61   See ICAO, *Report of the 29th Session of the ICAO Legal Committee,* ICAO Doc.9630 – LC/189 (1994) [hereinafter *Report of the 29th Session*], *Checklist of Items to be Considered in Contracts for GNSS Signal Provision With Signal Providers in the Context of Long-term GNSS* at para.3:71.2.

62   The terms of reference of the Panel of Experts on the Establishment of a Legal Framework With Regard to Global Navigation Satellite Systems established by the Council on 6 December 1995 comprise the preparation of draft texts, including a model contract, using the checklist approved by the 29th Session of the Legal Committee, for consideration by the ICAO Council. See ICAO, *Report of the Panel of Experts on the Establishment of a Legal Framework with regard to GNSS,* ICAO Doc.LTEP/1 (23 December 1996) [unpublished][hereinafter *LTEP/1 Report*].

63   See LTEP/1-WP/5, *supra* note 60 at paras.2.1 and 2.2.

64   See *ibid.* at para.2.2.4.

65   For more on the issue of channelling of liability, see below, Chapter 6 at 105 and Chapter 7 at 139.

66   See ICAO, *Statement of ICAO Policy on CNS/ATM Systems Implementation and Operation,* ICAO Doc. LC/29 - WP/3-2 (28 March 1994) [hereinafter *Council Statement*].

67   See Kotaite, *supra* note 60 at 198; Henaku, *supra* note 52 at 86-88; Milde, *supra* note 40 at 200.

68   See *AN-CONF/10 Report, supra* note 58, Recommendation 4/4 at 4.7.

69 *Ibid.*, Supplement No. 1 at 4.
70 See ICAO, *Legal Committee, 29th Session*, Report of the Rapporteur on the "Consideration, with regard to global navigation satellite systems (GNSS), of the establishment of a legal framework", by Kenneth Rattray, LC/29-WP/3-1 (3 March 1994) at 6 [hereinafter *Rattray's Report*].
71 See *Rattray's Report, ibid.*, Annex III.
72 *Ibid.* at 7.
73 *Report of the 29th Session, supra* note 61 at para.3:38.7.2.
74 See *Draft Agreement Between the International Civil Aviation Organisation (ICAO) and GNSS Signal Provider Regarding the Provision of Signals For GNSS Services,* reproduced in ICAO Doc.9630-LC/189 (1984) at para.3:38.10.
75 See Letters, *supra* note 39.
76 See *ibid.* See also L.Weber and A.Jakob, "Activities of the International Civil Aviation Organisation" (1996) XXI:II *Annals of Air and Space Law* 403 at 407.
77 B.D.K. Henaku, "The International Liability of the GNSS Space Segment Provider" (1996) XXIII:I *Annals of Air and Space Law* 145 at 155-156 [hereinafter Henaku].
78 See *Vienna Convention on the Law of the Treaties*, 23 May 1969, 1155 U.N.T.S. 331, Section 2, Articles 19-23 (entered into force 27 January 1980)[hereinafter *Vienna Convention*].
79 See Henaku, *supra* note 77 at 155-156.
80 Kotaite, *supra* note 60 at 203.
81 Kotaite, *ibid.*
82 *Report of the 29th Session, supra* note 61 at para.3:28.
83 See *ibid.* at para.3:31.
84 See Milde, *supra* note 40 at 201.
85 *Chicago Convention, supra* note 6, art.65.
86 Milde, *supra* note 40 at 201.
87 H.G.Schermers and N.M.Blokker, *International Institutional Law: Unity Within Diversity*, 3rd ed. (The Hague: Nijhoff, 1995) at 1096.
88 Henaku, *supra* note 52 at 182.
89 Henaku, *ibid.* at 185.
90 *Ibid.*
91 I.Brownlie, *Principles of Public International Law* (Oxford: Clarendon Press, 1998) at 643 [hereinafter Brownlie]. The author cites the *Nuclear Tests* Case (Australia v. France), whereby the ICJ held that "France was legally bound by publicly given undertakings, made on behalf of the French government, to cease the conduct of atmospheric nuclear tests. The criteria of obligation were: the intention of the State making the declaration that it should be bound according to its terms; and that the undertaking be given publicly".
92 Brownlie, *ibid.* at 644.
93 LTEP-WG/II, *supra* note 32, "Analysis of Liability Provisions in Existing International Conventions, Treaties and Other Relevant Instruments and Their Applicability to GNSS", LTEP-WG/II-WP/9 (18 April 1997), presented by O.Carel, M,Denney, E.Hoffstee, P.O'Neill, T.Nordeng. W. t'Hoen, A.Watt, G.White.
94 See *ibid.* See also J.M.Epstein, "Global Positioning System (GPS): Defining the Legal Issues of Its Expanding Civil Use" (1995) 61 *Journal of Air Law and Commerce* 243 at 276 [hereinafter Epstein].

95 U.S., *Treaties and Other International Agreements: The Role of the U.S. Senate, A Study Prepared for the Committee on Foreign Relations* (U.S. Senate, 103d Cong., 1st Sess., Nov. 1993) at xxxvii-xxxviii [hereinafter U.S. Senate Study].

96 Epstein, *supra* note 94 at 275.

97 U.S. Senate Study, *supra* note 95 at xvi.

98 Henaku makes particular reference to the Vienna Convention which states, in its Article 13, that the consent of a State to be bound by a treaty constituted by instruments exchanged between them is expressed by that exchange when: a) the instrument provides that their exchange shall have such effect; or b) it is otherwise established that those States were agreed that the exchange of instruments shall have that effect. See Henaku, *supra* note 52 at 183.

99 Letters, *supra* note 39.

100 See B. Cheng, *General Principles of Law as Applied by the International Courts and Tribunals* (Cambridge: Grotius Publications, 1987) at 118.

101 "Good faith in contractual relations implies the observance by the parties of a certain standard of fair dealing, sincerity, honesty, loyalty, in short, or morality, throughout their dealings". *Ibid.*

102 *LTEP/1 Report, supra* note 62.

103 See *Report of the 29ᵗʰ Session, supra* note 61 at para.3:29.

104 See *LTEP/1 Report, supra* note 62, Appendix 3 at A3-1.

105 See *ibid.*, Appendix 4 at A4-1. A review of the mandate given to Working Group II and the results of its work will follow below.

106 See ICAO, Assembly, 32ⁿᵈ Session, CD-ROM (Montreal, 1998), Charter on the Rights and Obligations of States Relating to GNSS Services, Res. A-32-19 [hereinafter Charter].

107 See ICAO, *Report of the Panel of Legal and Technical Experts on the Establishment of a Legal Framework with regard to GNSS,* ICAO Doc.LTEP/2 (3 November 1997) at paras.1:73 - 1:83 [unpublished][hereinafter *LTEP/2 Report*].

108 See ICAO, *Panel of Legal Experts on the Establishment of a Legal Framework with regard to GNSS, Working Group on GNSS Principles (Working Group I),* LTEP-WG/I (10-14 March 1997) [hereinafter LTEP-WG/I], "Introductory Note", LTEP-WG/I-WP/2 (20 February 1997).

109 See *LTEP/2 Report, supra* note 107 at para.1:87.

110 *LTEP/2 Report, ibid.* at para.1:88.

111 See *ibid.* at paras.1:89-1:91.

112 ICAO, *Assembly, 32ⁿᵈ Session, Legal Commission,* "Progress in the Work of the Panel of Legal and Technical Experts on the Establishment of a Legal Framework with Regard to GNSS (LTEP)", ICAO A-32-WP/24, LC/3 (18 June 1998).

113 ICAO, *World-wide CNS/ATM Systems Implementation Conference Report,* ICAO Doc.9719 (May 1998), Conclusion 5.2 at para.5.2.1[hereinafter *WW/IMP Report*].

114 See *Charter, supra* note 106.

115 See Milde, *supra* note 56 at 209.

116 See LTEP/1-WP/5, *supra* note 60 at para.5.2; *WW/IMP Report, supra* note 113 at para.5.1.5.

117 See LTEP/1-WP/5, *ibid.* at para.5.1.

118 *LTEP/1 Report, supra* note 62, Appendix 4 at A4-1.

119 See generally, ICAO, *WG/ II Report, supra* note 44. See also *LTEP/2 Report, supra* note 107 at para.2:2.

120 See ICAO, *Panel of Experts on the Establishment of a Legal Framework with regard to GNSS, Working Group on GNSS Framework Provisions (Working Group II), Second Meeting,* LTEP-WG/II(2) (2-5 September 1997) [hereinafter LTEP-WG/II(2)], "Report of the Results of the Informal Survey Conducted by Working Group II", LTEP-WG/II(2)-WP/2 (14 August 1997).

121 See *LTEP/2 Report, supra* note 107 at para.2:3.

122 ICAO, *Report of the Second Meeting of the Working Group on GNSS Framework Provisions (Working Group II) of the Panel of Legal and Technical Experts on the Establishment of a Legal Framework with Regard to GNSS (LTEP)* (5 September 1997), ICAO LTEP/2-WP/4 (15 September 1997) [unpublished]; ICAO, *Report of the Third Meeting of the Working Group on GNSS Framework Provisions (Working Group II) of the Panel of Legal and Technical Experts on the Establishment of a Legal Framework with Regard to GNSS (LTEP)* (12 February 1998), Appendix 3 to LTEP/3 Report [unpublished].

123 See ICAO, *Report of the Panel of Legal and Technical Experts on the Establishment of a Legal Framework with regard to GNSS*], ICAO Doc.LTEP/3 (9 March 1998) at para.1:1 [unpublished][hereinafter *LTEP/3 Report*].

124 See *ibid.* at para.1:41.

125 See ICAO, *Assembly, 32nd Session, Legal Commission, Recommendations of LTEP*, ICAO Doc.A-32-WP/24, Appendix B [hereinafter *LTEP Recommendations*]. An analysis of these legal aspects with a view to the long-term framework for GNSS will be provided in Chapter 7.

126 See *ibid.* at para.2.3.

127 See *ICAO Secretariat*, Transition, ICAO CNS/ATM Newsletter 98/5, "Rio Lays Institutional and Financial Groundwork" (Autumn 1998). The Conference was attended by participants from 123 Contracting States, 27 international organisations and 38 industry delegations. See *WW/IMP Report, supra* note 133 at para.2. See above at 15.

128 See *WW/IMP Report, ibid.* at para.3.1.1.

129 See *ICAO Secretariat*, Transition, ICAO CNS/ATM Newsletter 98/5, "Significant majority of States need Help" (Autumn 1998).

130 See *ibid.* at paras.2.1 and 2.2.

131 *WW/IMP Report, supra* note 113, Recommendation 5/3 at para.5.3.1.

132 See *ibid.,* Recommendation 5/4 at para.5.3.1.

133 See *ibid.*, Conclusion 5/2 at para.5.2.1.

134 ICAO, Assembly, 32nd Session, CD-ROM (Montreal, 1998), Development and Elaboration of an Appropriate Long-term Legal Framework to Govern the Implementation of GNSS, Res. A-32-20.

135 See *WW/IMP Report, supra* note 113, *Declaration on Global Air Navigation Systems for the Twenty-first Century,* at para.7.2 [hereinafter *Rio Declaration*].

6 The Long-Term Legal Framework

Forms of Instrument

Since the item "consideration, with regard to global navigation satellite systems (GNSS), of the establishment of a legal framework" was given highest priority in the General Work Programme of the Legal Committee in 1992, there have been extensive discussions on the form and content of such a legal framework.

In particular, at the 29th Session of the Legal Committee, following a proposal by the Rapporteur on the subject, Dr. Kenneth Rattray, whereby it was submitted that the legal framework should be established by an international convention or agreement under the auspices of ICAO,[1] questions arose as to the need or even desirability of the elaboration of an international legal instrument. Opinions differed, as States with the greatest institutional concerns would favour an international convention, in opposition to the *de facto* signal provider States. Most delegations, however, due to inherent delays in drawing up, adopting and bringing into force an international legal instrument, and bearing in mind the urgency of the task, favoured the adoption of a "step-by-step approach".[2]

In this respect, it should be reiterated here that a consensus was reached on a two-stage approach concerning the development of a legal framework for the existing systems and the elaboration of a more complete and lasting instrument for the future.[3]

Work conducted by the LTEP with a view to the long-term GNSS has identified various private and public law options and considered the pros and cons for different types and forms of legal framework, namely: i) checklist; ii)

101

model contract; iii) codes of conduct, guidelines and guidance material; iv) standards and recommended practices; v) Assembly or Council resolution, or Assembly declaration; vi) international agreement, or international convention; or vii) a combination.[4] A practical interpretation of these options has already been given in the previous chapter with an emphasis on the already existing international arrangements.

It has been agreed that the complexity of the legal framework for GNSS would require "a combination of various types and forms since one could not possibly conceive that a single instrument would provide a complete legal framework".[5]

Need or Desirability of an International Convention

The years have seen the development of two clear schools of thought on the need of a new legal framework and the desirability of an international convention to govern the implementation and operation of GNSS. The matter has been subject to lengthy discussions under the aegis of ICAO and other international fora. Discerning views recently expressed by three distinguished delegates and eminent speakers at the World-wide CNS/ATM Systems Implementation Conference, as well as remarks made by one legal expert at Eurocontrol, will serve here as a basis for discriminating between the opposing arguments and the different perspectives of the international community.

The Signal Providers' Perspective

"GNSS not only has a legal framework, it has a framework which is adequate to the task",[6] has argued Michael B. Jennison, Assistant Chief Counsel for International Affairs of the U.S. Federal Aviation Administration.

Accordingly, in his opinion, the existing legal framework, namely the Chicago Convention, is flexible enough and has managed to adapt over the years to bigger technical developments than satellite navigation, including the development of increasingly sophisticated navaids. In this respect, he argues that "despite the apparent quantum leap in capabilities, satellite navigation, in the legal and institutional issues that it presents, has had its precursors in both short-range and long-range navigation systems in use for many years ... across national boundaries".[7] And goes further to remind that such navaids have also been first developed by the military, having gradually come under civil control.

In his view, the Chicago Convention "*is* the legal framework for CNS/ATM",[8] a legal framework which already comes along with the necessary mechanism to be kept updated, through the adoption of SARPs:

> We have the basic, constitutional law that we need (the kind that takes many years to develop and to bring into effect), and we have the ready means to make additional law - both *binding* rules and *non binding* guidance – to fill in any significant gaps that might emerge. (And no significant gap has emerged so far.)[9]

As far as liability is concerned, Mr. Jennison purports to demonstrate that the absence of a multilateral liability regime for air traffic control agencies has never meant there are no liability rules or that people are barred from pursuing remedies in courts of law, but simply that "there has been no demonstrated practical need".[10] Similarly, legal channels do exist for liability claims with respect to satellite navigation.

With respect to the long-term legal framework, the development of which he considers to be an expensive, impractical and time-consuming solution, and recalling that GNSS has been declared to be fully consistent with the Chicago Convention, he beseeches "legal parsimony", and his conclusions could be summarised with the following assertion:

> Only when we have a clear conception of what may constitute the long-term GNSS, can lawyers and policy makers then contemplate whether additional law, in whatever form, may be needed. Indeed the development of the law *typically* follows social, economic, and technical developments.[11]

The User States' Perspective

The views expressed above meet with strong opposition in other areas of the world, particularly Europe and developing countries, whose concerns are represented here in the following dissenting words of Dr. Kenneth Rattray, Solicitor General of Jamaica:

> The simple assertion that GNSS is not inconsistent with the Chicago Convention provides no assurance or comfort for the implementation of the system with integrity. The principles contained in the International Air Law Conventions ... are all compatible with the Chicago Convention. But compatibility of principles has never been regarded as an adequate basis for engaging the responsibility of States. A Convention is necessary for this purpose. In the field of GNSS it is more so necessary because many of the elements of the system will be outside the sovereign control of States.[12]

In his opinion, reality speaks that "the GNSS facilities, at least as far as the space segment is concerned, will be controlled and operated by one or more foreign countries representing a dramatic step away from past practice in the provision of air navigation services".[13] Consequently, not only does it pose a challenge to the application of the principle of sovereignty but also to the confidence in the integrity of the system and political influences which might undermine its credibility.[14] In this sense, he considers the goodwill or good faith of the signal provider States an inadequate basis on which to build the required confidence.[15] Before authorising the use of GNSS signals in their sovereign territory, States want certain guarantees to be provided in the form of an international convention, including the proper allocation of liabilities.

Bearing in mind the importance of financing for the implementation and operation of the systems, he recalls that "most administrations, especially in developing countries, will have to make significant economic and budgetary decisions regarding the aviation communication and navigation infrastructure in the short, medium and long-term".[16] Financial security and system credibility in the system are therefore essential to allow for a decision to be made as to whether contemplated investments in INS or ILS should be abandoned for a quantum leap into GNSS.[17]

Counter to a view previously held[18] that the market place would ultimately determine when GNSS would be accepted and the degree to which it would be relied upon, he considers it an inadequate mechanism to provide the necessary guarantees which will inspire world-wide confidence:[19]

> It is absolutely essential that the foundation of GNSS on a world-wide basis be construed on pillars of political confidence, pillars of financial confidence and pillars of technical and technological confidence. The three pillars must be anchored and secured by legal and institutional foundations which can only be provided by an international convention which spells in detail the fundamental principles governing the implementation of GNSS.[20]

Calling for the necessary assurances as regards universal accessibility, reliability, continuity, affordability, liability, international co-operation and oversight by ICAO with a view to the long-term legal framework, he concludes that the adoption of the Charter in the form of an Assembly resolution, as an interim measure, could be regarded as a "first step" towards an international convention[21] establishing fundamental principles, as well as legal obligations and respective liabilities to States and service providers.[22]

An Alternative View

An alternative view, previously outlined during LTEP discussions at the initiative of Eurocontrol,[23] advocating that while an international convention would be desirable for the long-term future, an interim approach could take the form of regional arrangements and a chain of contracts, was also endorsed by the Conference, and finds itself expressed here through the remarks then made by Mr. Trond.V.Nordeng, Managing Director at Nordic Aviation Resources S.A. (Norway). Accordingly:

> It should be envisaged that quite some time will pass before a long-term legal instrument will be adopted. ... The alternative is to establish a chain of contracts, firstly between primary signal provider and regional augmentation service provider; secondly between the... service provider and the States which have authorised the use of GNSS in their airspace which may also supply local augmentation service and equipment.[24]

The contractual chain approach could be described as a series of contractual arrangements between the various components of the system, where individual performance criteria would be established. While providing the necessary guarantees, it would clearly identify the extent of responsibility for the different actors at each stage of the chain. In case of an accident, channelling of liability would eventually trace it to the party whose actions or omissions had been the cause of the damage. Therefore, in case of joint and several liability, each actor would bear only the part of a global liability which corresponds to the extent that its action or omission contributed to the damage.[25]

The proposal of the Eurocontrol had used the term "regulatory chain" to describe such structure, which could be broken into four main elements with distinct roles, namely the signal-in-space provider, the augmentation provider, the user State and the end user, the latter being, in terms of aviation, the aircraft operator equipped with a GNSS receiver for navigational purposes. The roles of the user State could be further divided into safety regulation and air traffic services provision. Reference was especially made to a European GNSS Agency, which would undertake on behalf of States activities covering operational, technical, financial and institutional matters.[26]

In the same vein, Mr. Nortend contends that, in order to avoid a "prohibitive administrative burden" on individual States, an interface body, either an existing international organisation or a new GNSS entity, should be established and vested with the appropriate powers to negotiate regionally on behalf of States.[27]

The advantages of the contractual framework as an interim approach were at another occasion voiced by Mr. Roderick van Dam, Head of Legal Services at Eurocontrol, who stated that:

> The contractual chain approach has certain advantages as an interim solution. It allows for the specific requirements of a particular region to be reflected by agreements. It is well-suited for evolution. It offers flexibility, since new agreements can be drafted as new players join the system, without necessarily having to revise existing arrangements. It provides for speedy implementation if the parties are willing and ready to formalise their relationship. Finally, it offers a seamless path to the long-term resolution of an international convention.[28]

The Predominant View

It should be restated here that the predominant view at the Conference was that the adoption of a Charter was only one step in the long-term legal framework, which should take the form of an international convention. This view was widely supported by member-States of ASECNA,[29] ECAC[30] and LACAC,[31] as well as India and Nepal, the other States in the Asian region having not manifested their opinion in this regard. Contrary to the reference to an international convention in the conclusions of the Conference were, inter alia, the U.S., New Zealand, Australia and Canada.[32] ECAC, in particular, stands for a regional approach through the adoption of a chain of contracts among all the relevant actors.[33] The opposing views found their common denominator in that further work on the legal issues should not delay implementation of the systems.[34]

An Afterthought

The antithesis so vividly expressed in these presentations and ensuing discussions at the Conference in Rio de Janeiro is but a clear reflection of a race towards market dominance, where there is definitely no premium for a supporting role. When competing political and economic interests give battle, only the political will of States brings any likelihood of success. Whether an international convention will be the one to come along with the white flag of consensus[35] remains a possibility best envisaged for the long-term. Perhaps the more concrete regional approach of a contractual framework will best suit the discerning views and contribute to the development of a global legal framework through the comparison of regional solutions. Meanwhile, the implementation of CNS/ATM systems should not be delayed pending work on the legal issues.

Fundamental Principles

Safety of International Civil Aviation

Safety emerges as the primary principle in the regulation and standardisation of international civil aviation. The concept finds itself embodied in the Preamble and Article 44(h)[36] of the Convention. In addition thereto, standards and recommended practices covering a wide array of technical and operational regulation for world-wide application and essential to safe air navigation can be found in the Annexes to the Convention. The level of safety and efficiency of air transport is therefore directly linked to the uniform and effective implementation of SARPs.[37]

Specific reference is made in paragraph 1 of the Charter, whereby States recognise that in the provision and operation of GNSS services, the safety of international civil aviation shall be the paramount principle.

In fostering the development of a seamless, globally co-ordinated CNS/ATM system which aims to improve upon the overall efficiency of airspace and airport operations, leading to increased capacity and availability of user-preferred flight schedules and profiles, safety must remain the greatest concern. There can be no compromise. Safety in aviation is paramount and all other considerations are subordinate to it. In the words of ICAO's Secretary General, R.C. Costa Pereira:

> In the absence of safe and secure operations, aviation cannot drive economic and social development. Safety is the primary concern of the world aviation community, and identification of safety issues, funding and implementation of safety-related projects are foremost on ICAO's priority list.[38]

Global Aviation Safety Plan

ICAO continues to fulfil its primary objective, that of promoting the safety of international civil aviation world-wide. At the 32nd Session of the ICAO Assembly, a Resolution[39] was adopted endorsing the establishment of the ICAO Global Aviation Safety Plan (GASP), the main objective of which is to achieve a significant reduction in the number and rates of aviation accidents on a global basis. Acting as an umbrella document for all safety-related activities of ICAO, it comprises, *inter alia,* the following activities: i) recommend safety actions in response to findings of the Universal Safety Oversight Audit Programme, including rectification of deficiencies;[40] ii) improve the existing safety database systems so as to facilitate the dissemination of safety-related information between governments and industry;

and iii) enhance the identification of, an address all safety issues, such as shortcomings and deficiencies in the air navigation field.[41]

An element of the GASP, USOAP identifies such deficiencies and recommends remedial actions, whereas it is the responsibility of States to implement a corrective action plan and to ensure continuous compliance with ICAO SARPs. Many States, however, in spite of their best interests and efforts, are facing serious difficulties in fulfilling such tasks. Taking into account the significant impact that the lack of implementation or inadequate implementation of air navigation infrastructure, related safety standards and operation of air navigation services would have on safety world-wide, considerable effort is being made in identifying technical, financial and organisational corrective measures.[42]

On the technical side, and particularly relevant to this study, it is expected that, over the longer-term, CNS/ATM systems will effectively remedy many safety-related problems. Civil aviation, however, depends also on the continued availability of properly trained personnel to operate the system safely.[43] The importance of human factors and training issues in the implementation of the new systems have already been addressed in Chapter 1.

On the other hand, "financing" safety is one of the most critical factors for the implementation of the CNS/ATM system, including the required airport and air navigation services infrastructure. It is undeniable that under many developing States' present regulatory structures, budgetary allocations for civil aviation projects, including remedying of deficiencies are extremely limited due to other national expenditure priorities. As a result, addressing of identified shortcomings and deficiencies in air navigation services have continuously been left aside. The establishment of an International Financial Facility for Aviation Safety (IFFAS), which shall have as its over-riding priority to finance safety-related projects, will be discussed in Chapter 7.

Commenting on the issue of financing, Mr. Costa Pereira stated that "a particular challenge in many States is the lack of identification of all costs attributable to the provision and operation of airports and air navigation services, which remains an obstacle to full recovery".[44] Charge revenues need to be distributed to those entities actually providing the facilities for which the charges are levied. Application of a sound methodology for determining the cost basis for charges and effective collection mechanisms might be the solution.[45]

Hence, organisational structures under which airports and air navigation facilities are most effectively operated must ensure financial and operational autonomy. Privatisation is but one form of accomplishing that since ownership can rest in either public or private hands or a mixture of both. Additional benefits in the cost-effective implementation of CNS/ATM systems can be achieved through co-operative ventures between States, such as joint financing arrangements, international operating agencies, and joint charges collection agencies.[46] These institutional issues will receive further consideration in Chapter 7.

In brief, the need for close co-operation between States providing the signal in space, user States, airspace users, planning and implementation regional groups, governing bodies of ICAO and the Secretariat, at both global and regional levels, must be acknowledged so as to guarantee safety in the implementation of the CNS/ATM systems.

Universal Accessibility Without Discrimination

The primary precept in the Council Statement has its foundation in Article 15 of the Chicago Convention, which stipulates that uniform conditions shall apply to the use, by aircraft of every contracting State, of airports and air navigation facilities, including radio and meteorological services, which may be provided for public use for the safety and expedition of air navigation.[47]

It has also been reiterated in the Exchange of Letters, as well as in the Charter on the Rights and Obligations of States Relating to GNSS Services, which stipulates that:

> Every State and aircraft of all States shall have access, on a non-discriminatory basis under uniform conditions, to the use of GNSS services, including regional augmentation systems for aeronautical use within the area of coverage of such systems.[48]

The expression "under uniform conditions" employed therein emphasises the understanding that the principle does not imply the non-existence of rules or conditions of access, but simply that such rules or conditions must be equal to everyone.[49]

In practical terms, economic competition and the multiplicity of service providers have definitely been playing a categorical role in providing the necessary guarantees of accessibility in the provision of satellite communication services. At present, however, the same cannot be said to be true in the field of air navigation.[50]

Concerns have been raised about the dangers of a monopoly being exerted by the U.S. through the GPS, especially because GNSS users, namely, aircraft operators, Article 28 States, and the actual providers of air traffic services will be relying on a foreign system whose signals are generated outside their territory, and therefore is not directly under their control.[51]

A good example has been set by Inmarsat,[52] the major AMMS service provider, who has acknowledged the principle under Article 7 of its Convention.[53] In addition thereto, the contracts between Inmarsat and its signatories for the provision of transponder segment capacity for GNSS services on Inmarsat-III satellites contain a specific requirement that access shall be without discrimination on grounds of nationality or type of use.[54]

Views have been expressed that "States cannot make an important investment decision to change navigation systems on the basis of a fragile contractual and commercial relationship that can be changed at any time".[55] Hence, an international convention would be the best means for providing this legal guarantee. Furthermore, it has been submitted that SARPs, traditionally used to regulate technical and operational matters, are not the appropriate instrument for dealing with the issue of universal accessibility.[56]

Reliability and Continuity of the Services

Continuity of a system has been defined as the capability of the system to perform its function without non-scheduled interruption during the intended operation.[57] In a wider legal sense, it has been referred to as the principle that the services are not to be interrupted, modified, altered or terminated for military, budgetary or other non-technical reasons.[58]

In the preceding chapter, the analysis of the institutional aspects regarding the use of GNSS as a sole means of navigation confirmed there are a number of factors that might influence the performance of GNSS, all of which raise important concerns with respect to the sole reliance on the services provided. Although it has been established that the provision of GNSS services will always follow the principle of redundancy,[59] with options ranging from an automatic switch to a back-up system on stand-by to an institutional guarantee by an international organisation, legal guarantees as to the technical performance of the system have judiciously been demanded by the international community.

The letters exchanged between ICAO and the service providers have both recognised the principle, which has also been incorporated in the Council Statement. More recently, the Charter has stipulated that:

Every State providing GNSS services, including signals, or under whose jurisdiction such services are provided, shall ensure the continuity, availability, integrity, accuracy and reliability of such services, including effective arrangements to *minimise the operational impact of system malfunctions or failure, and to achieve expeditious service recovery*. Such State shall ensure that the services are in accordance with ICAO Standards. States shall provide in due time aeronautical information on any modification of the GNSS services that may affect the provision of the services.[60]

Nevertheless, in a pragmatic point of view, it has been submitted that no such international guarantees will ever be obtained from the current GNSS service providers. In the words of Michael Milde, "no better guarantees can be assured even by a purely civilian GNSS under international control; even that system would be vulnerable to an act of God, any international crisis, or simply a lack of funds".[61]

The discontinuation of services as a possibility out of Article 89 of the Chicago Convention has also been considered by the LTEP. The article appears to be a "legal loophole"[62] in the sense it clearly states that freedom of action is reserved to States in case of war or national emergency. A few have expressed their dissenting opinion by stating that such freedom is not unlimited and that once the possibility of interrupting services for non-technical reasons is recognised as acceptable, the system will no longer fulfil the requirements for a global air navigation system.[63] Having already elevated safety of international civil aviation to the first principle in the Charter, Working Group I decided that it was not the proper forum to discuss the legal implications of Article 89, which involved complex questions of international law, regarding armed conflict, including rights and obligations of belligerents.[64]

Sovereignty of States

In accordance with customary international law, the competence of States in respect to their territory and its appurtenances, namely airspace and territorial sea, together with the government and population within its frontiers is described in terms of State sovereignty and jurisdiction.[65] Recognising the complete and exclusive sovereignty of every State over the airspace above its territory, Article 1 of the Chicago Convention is therefore merely declaratory in nature.[66] For this purpose, the territory of a State is deemed to be "the land areas and territorial waters[67] adjacent thereto under the sovereignty, suzerainty, protection or mandate of such State".[68]

As a corollary, there is no "freedom of the air" above a State's territory. No scheduled international air service may be operated over or into the territory of a contracting State, except with the special permission or authorisation of that State, and in accordance with such terms.[69]

Additionally, upon entering or departing a State's territory, and while within, any aircraft engaged in international air navigation shall comply with that State's laws and regulations relating to the operation and navigation of such aircraft.[70] Likewise, in a GNSS environment, it has been declared that the authority and responsibility of a State to control operations of aircraft within its sovereign airspace shall be preserved, having these rights been expressly recognised by the Charter.[71]

The responsibility of States under Article 28 to provide, as far as they may find practicable, air navigation facilities and services in their territory[72] flows from their sovereignty.[73] As will be seen later in this Chapter, in the exercise of sovereignty, States may delegate the actual technical operation of these services to a third party, although remaining ultimately responsible for the provision and regulation of the services.[74]

The fact that GNSS facilities, particularly the space segment, will be operated and controlled by one or more foreign States has been said to represent "a dramatic step away from past practice in the application of the principle of sovereignty",[75] when States retained full control of all the elements of the services provided, and therefore were fully responsible to ensure their compliance with ICAO SARPs. Concerns have especially been raised as to the time when GNSS is approved as the sole means of navigation and the whole navigation system will be outside the territorial control of these States who undertook responsibility under Article 28.[76]

In response to such concerns, the sovereign rights of a State to regulate and control air navigation services within its territory, in the event they decide to avail themselves of the GNSS signals as an aid to navigation, have been duly acknowledged in the guidelines, the Council Statement as well as the Exchange of Letters, and also reaffirmed by the Charter. Accordingly:

> The implementation and operation of CNS/ATM systems shall neither infringe nor impose restrictions upon States' sovereignty, authority or responsibility in the control of air navigation and the promulgation and enforcement of safety regulations. State's authority shall be preserved in the co-ordination and control of communications and in the augmentation, as necessary, of satellite-based navigation services.[77]

On the other hand, views to the contrary have expressed that a State providing space segments for GNSS only provides signals enabling positioning and navigation of aircraft which cannot be deemed to be services within the meaning of Article 28. Thus, in their opinion, the very technical and passive nature of GNSS should be considered a sufficient safeguard that the above requirements will be met.[78]

From a pragmatic point of view, the absence of any reference in the Chicago Convention to any specific level of facilities and services to be provided in the sovereign territory of a State is interpreted to mean that no State is actually obliged to make use of satellite technology as an aid to air navigation and air traffic control, having to specifically authorise the use of the signal-in-space in its airspace[79] and satisfy itself that it complies with ICAO SARPs. In this respect, it has been submitted that:

> GNSS cannot and will not be imposed on States against their will and their support of the GNSS will depend entirely on their sovereign political will. Nevertheless, the full benefits of the GNSS will be available only to those States that will accept an agreed co-operative framework for the GNSS.[80]

In brief, it is essential that a compromise be reached between the need to ensure the effective global use of CNS/ATM technology in a seamless airspace, where territorial State boundaries and Flight Information Regions cease to be of primary importance, and the need to respect State sovereignty.[81] Some flexibility in the exercise of sovereign rights might therefore be necessary, in particular in the delegation of tasks of signal provision and augmentation to foreign States and/or joint agencies or operating structures.[82]

Co-operation and Mutual Assistance

Co-operation and mutual assistance have been deemed essential in the planning, implementation and operation of the CNS/ATM systems, gaining special consideration in view of ICAO's objective to achieve a seamless, interoperable and global system.[83] The Charter acknowledges this principle in its paragraph 7, having also provided that every State shall conduct its GNSS activities with due regard for the interests of other States.

Results of two detailed technical surveys carried out by ICAO in 1994 and 1997 revealed that a majority of States requires external assistance. The broad areas of concern range from "a formal needs assessment survey, through implementation planning, including cost/benefit analysis and system procurement, to human resources planning and development".[84] The surveys have also indicated the preference of States for such assistance to be provided by ICAO.

For example, financing of technical co-operation will require unprecedented co-operative efforts on the part of the States and the international developing financing community alike.[85] It has been submitted that developed States and other donors should assist States experiencing difficulties in obtaining funding for CNS/ATM planning and implementation. Furthermore, ICAO should continue its important co-ordinating role of technical co-operation in close consultation with all partners in the systems.[86]

The Role of ICAO

The central role to be played by ICAO, as the international organisation in the best position to effectively monitor and co-ordinate the planning and implementation of the CNS/ATM systems must be recognised by States. In particular, five main functions are to be performed at the regional and global levels, as envisaged by the LTEP recommendations and affirmed by the Council Statement:

a)   responsibility for the establishment of appropriate standards, recommended practices and procedures, in accordance with Article 37 of the Chicago Convention;

b)   global co-ordination and monitoring of the systems on a global basis in accordance with the global co-ordinated CNS/ATM systems plan and the regional air navigation plans to ensure compatibility and interoperability of the different systems;

c)   facilitation of assistance to States with regard to the technical, financial, managerial, legal and co-operative aspects of the systems' implementation;

d)   co-ordination with other international organisations in any matter related to GNSS, including the use of the frequency spectrum in support of international civil aviation;[87]

e)   any other related function within the framework of the Chicago Convention, including those under Chapter XV of the Convention.[88]

It has been stated that the fact that both the U.S. and the Russian Federation have provided their respective systems for use by the international community through the forum of ICAO is a clear indication of their recognition of the organisation's central role in the planning and implementation of the CNS/ATM systems.[89]

Compatibility of Regional Arrangements with Global Planning and Implementation

Particularly important in the context of global co-ordination is the need for States to ensure that regional or sub-regional arrangements are not only compatible with the global planning and implementation process of GNSS,[90] but also a means to promote the integration of the system. Therefore systems are to be devised and implemented according to a well prepared plan, and full co-operation is required at the international level so as to provide for the optimum use of the limited financial resources, minimise duplication of efforts, and prevent mutual interference.[91]

Notes

1   See ICAO, *Legal Committee, 29th Session*, Report of the Rapporteur on the "Consideration, with regard to global navigation satellite systems (GNSS), of the establishment of a legal framework", by Kenneth Rattray, LC/29-WP/3-1 (3 March 1994) at paras.9 and 18.

2   See ICAO, *Report of the 29th Session of the ICAO Legal Committee*, ICAO Doc.9630 – LC/189 (1994) at para.3:28 [hereinafter *Report of the 29th Session*].

3   See *Report of the 29th Session, ibid.* at para.3:29. See also J.Huang, "Development of the Long-Term Legal Framework for the Global Navigation Satellite System" (1997) XXII:I *Annals of Air and Space Law* 585 at 586-587 [hereinafter Huang].

4   See ICAO, *Panel of Experts on the Establishment of a Legal Framework With Regard to GNSS*, LTEP/1 (25-30 November 1996) [hereinafter LTEP/1], "Different Types and Forms of the Long-Term Legal Framework For GNSS", LTEP/1-WP/5 (20 September 1996).

5   ICAO, *Report of the Panel of Experts on the Establishment of a Legal Framework with regard to GNSS,* ICAO Doc.LTEP/1 (23 December 1996) at para.4:1.14 [unpublished][hereinafter *LTEP/1 Report*].

6   M.B.Jennison, "A Legal Framework for CNS/ATM Systems" (ICAO World-wide CNS/ATM Systems Implementation Conference, Rio de Janeiro, 14 May 1998) [hereinafter Jennison] at 1. See also M.B.Jennison, "The International Law of Satellite-Based Navigation Aids" (American Bar Association Forum on Air and Space Law, Montreal, 3 August 2000).

7   Jennison, *ibid.* at 1-2.

8   *Ibid.* at 2.

9   *Ibid.* at 1.

10   *Ibid.* at 6.

11   *Ibid.* at 5.

12   K.O.Rattray, "Legal and Institutional Challenges for GNSS – The Need for Fundamental Obligatory Norms" (ICAO World-wide CNS/ATM Systems Implementation Conference, Rio de Janeiro, 14 May 1998) at 7 [hereinafter Rattray].

13   Rattray, *ibid.* at 4.

14   See *ibid.*

15   See *ibid.* at 7.

16   *Ibid.* at 5.

17   *Ibid.*

18   See ICAO, *World-wide CNS/ATM Systems Implementation Conference Report*, ICAO Doc.9719 (May 1998) at 5.1.4 [hereinafter *WW/IMP Report*].

19   See Rattray, *supra* note 12 at 7-8.

20   *Ibid.* at 1.

21   *Ibid.* at 8.

22   See *ibid.* at 6.

23   See LTEP/1, *supra* note 4, "Outline of the Role and Functions of a Multi-Modal European GNSS Agency and its Place Within the Regulatory Chain", ICAO LTEP/1-WP/16 (25 November 1996) at para.4 ff., presented by Eurocontrol [hereinafter LTEP/1-WP/16].

24 T.V.Nordeng, "International Legal Impact on National Implementation of Global Navigation Satellite Systems (GNSS)" (ICAO World-wide CNS/ATM Systems Implementation Conference, Rio de Janeiro, 14 May 1998) at 2 [hereinafter Nordeng].

25 See ICAO, *Panel of Experts on the Establishment of a Legal Framework With Regard to GNSS*, LTEP/2 (6-10 October 1997) [hereinafter LTEP/2], "Liability Aspects of GNSS", ICAO Doc.LTEP/2-WP/6 (1 October 1997) at para.5ff., presented by O.Carel, P.O'Neill, F.Schubert, R.D.van Dam, G.White, F.A.Wister. See also Huang, *supra* note 3 at 594; R.D.van Dam, "Recent Developments at the European Organisation for the Safety of Air Navigation (EUROCONTROL)" (1998) XXIII *Annals of Air and Space Law* 311 at 319 [hereinafter van Dam].

26 See LTEP/1-WP/16, *supra* note 23.

27 See Nordeng, *supra* note 24.

28 van Dam, *supra* note 25 at 319.

29 Agence pour la Sécurité de la Navigation Aérienne (Africa and Madagascar) (ASECNA).

30 European Civil Aviation Conference (ECAC).

31 Latin American Civil Aviation Commission (LACAC).

32 See I.Lagarrigue and J.D.Bloch, "Le GNSS et Le Droit des États: l'Affrontement Entre États Fournisseurs et États Utilisateurs Lors de la Conférénce de Rio sur le CNS/ATM" (1998) 43:183 Revue Navigation 345 at 347-348. See also, J.Dupont, "Une Convention Internationale pour le GNSS" (1998) 36:1661 *Air and Cosmos Aviation International*.

33 See van Dam, *supra* note 25 at 318.

34 See *WW/IMP Report*, *supra* note 18 at para.5.1.8. See *ibid.*, Recommendation 5/3.

35 For a very interesting review on the role of treaties as a medium for law-making in the contemporary world, see C.Lim and O.Elias, "The Role of Treaties in the Contemporary International Legal Order" (1997) 66 *Nordic Journal of International Law* 1 at 1-21, where the author states that "there is nothing inherent in the nature of the treaty system which singles it out as *the* vehicle for making the ideal of an international law of co-operation a reality" and that "the less initial common ground there is for a generally acceptable instrument to arise, the less likely it is that a treaty, or at least a useful treaty will come into existence". *Ibid.* at 14.

36 See Introduction, note 40 and accompanying text.

37 For further discussion on the legal significance of the ICAO SARPs, see Chapter 5, above at 80.

38 R.C.Costa Pereira, "Funding and Implementing Regional and Sub-regional Solutions in Africa" (African Aviation Conference and Exhibition 1999, Washington, 28 June 1999) [unpublished] [hereinafter Costa Pereira].

39 See ICAO, Assembly, 32nd Session, CD-ROM (Montreal, 1998), *ICAO Global Aviation Safety Plan (GASP)*, Res. A32-15 at paras.5, 9.

40 See Chapter 5 at 76.

41 See ICAO, *Air Navigation Commission*, "Approval of a Draft Assembly Working Paper - Progress Report on the ICAO Global Aviation Safety Plan", ICAO AN-WP/7626 (20 February 2001). See especially ICAO, *Council, 163rd Session*, "Draft Assembly Working Paper - Mobilisation of Funds for Civil Aviation", ICAO C-WP/11591 (8 May 2001) at para.4.2.

42   See ICAO, *Assembly, 32nd Session, Executive Committee*, "Shortcomings and Deficiencies in the Air Navigation Field", ICAO A-32-WP/96, EX-41, Appendix (13 August 1998) at 1 [hereinafter GASP].

43   See *ibid.* at 2.

44   Costa Pereira, *supra* note 38 at 8-9.

45   See *ibid.* at 9.

46   See GASP, *supra* note 42 at 2.

47   See *Convention on International Civil Aviation*, 7 December 1944, ICAO Doc.7300/6; UN Doc.15 U.N.T.S.295, art. 15 (entered into force 4 April 1947) [hereinafter *Chicago Convention*][emphasis added].

48   ICAO, Assembly, 32nd Session, CD-ROM (Montreal, 1998), Charter on the Rights and Obligations of States Relating to GNSS Services, Res. A-32-19 at para.2 [hereinafter Charter].

49   See ICAO, *Report of the Panel of Legal and Technical Experts on the Establishment of a Legal Framework with regard to GNSS*, ICAO Doc.LTEP/2 (3 November 1997) at para.1:29 [unpublished][hereinafter *LTEP/2 Report*]. The expression was incorporated as a result of a view expressed that the retention of the previous wording "without discrimination of any kind" could lead to a situation where a commercial provider would be obliged to provide services to all States regardless of whether the parties had reached an agreement or regardless of payment. *Ibid.* at para.1:28. Another point was raised as to whether in situations involving the imposition of United Nations sanctions discrimination would be considered justified. The Chairman, Mr. Gilles Lauzon (Canada), explained that according to the International Court of Justice, decisions of the Security Council are superior to treaty obligations. Therefore, it should be expected that any sanctions involving enforcement measures would be a Security Council decision. *Ibid.* at paras.1:28, 1:29.

50   See ICAO, *Global Air Navigation Plan for CNS/ATM Systems,* version 1 (Montreal: ICAO, 1998), vol. 1 at para.11.2.3.2 [hereinafter *Global Plan*]; Huang, *supra* note 3 at 588.

51   See *Global Plan, ibid.*

52   For information on the privatisation process of Inmarsat, see D. Sagar, "Recent Developments at the International Mobile Satellite Organisation (INMARSAT)" (1998) XXIII *Annals of Air and Space Law* 343 at 343-347.

53   See *Convention on the International Maritime Satellite Organisation* (INMARSAT) 3 September 1976, 1143 U.N.T.S. 105 (entered into force 16 July 1979, Article 7[hereinafter *Inmarsat Convention*].

54   See LTEP/1, *supra* note 4, "Inmarsat Satellite Navigation Services Institutional and Contractual Aspects", ICAO Doc.LTEP/1-WP/11 (29 October 1996) at para.3 [hereinafter LTEP/1-WP/11].

55   A.Kotaite, ICAO's Role with Respect to the Institutional Arrangements and Legal Framework of Global Navigation Satellite System (GNSS) Planning and Implementation (1996) XXI:II *Annals of Air and Space Law* 195 at 200 [hereinafter Kotaite].

56   See Kotaite , *ibid.* See also Rattray, *supra* note 12 at 4.

57   See ICAO, *Report on the Third Meeting of the Global Navigation Satellite System Panel*, Appendix C to the Report on Agenda Item 1, GNSS/3-WP/66 (12-23 April 1999) at para.C..3.4.1. [unpublished].

58 See *LTEP/1 Report, supra* note 5 at para.3.5.
59 See *Global Plan, supra* note 50, vol. I at para.11.2.4.
60 *Charter, supra* note 48 at para.4 [emphasis added].
61 M.Milde. "Solutions in Search of a Problem? Legal Aspects of the GNSS" (1997) XXII:II *Annals of Air and Space Law* 195 at 207 [hereinafter Milde].
62 *Ibid.*
63 See *LTEP/1 Report, supra* note 5 at paras.3:7-3:11.
64 See *LTEP/2 Report, supra* note 49 at para.1:41.
65 See I.Brownlie, *Principles of Public International Law* (Oxford: Clarendon Press, 1998) at 106. See also J.F.Rezek, *Direito Internacional Público* (São Paulo: Saraiva, 1996) at 163-164 [hereinafter Rezek].
66 See N.M.Matte, *Treatise on Air-Aeronautical Law* (Montreal: McGill University, 1981) at 132 [hereinafter Matte].
67 Contrary to the Law of the Sea, there is no right of innocent passage for aircraft over the territorial waters of a State. See I.H.Ph. Diederiks-Verschoor, *An introduction to Air Law*, 5th rev. ed. (Deventer: Kluewer Law and Taxation, 1993) at 30.
68 *Chicago Convention, supra* note 47, art. 2. This definition has given rise to a number of questions on how the terms of said article are to be applied, as regards the extension of the territorial waters and the setting up of exclusive economic zones. For more on the subject, see Matte, *supra* note 66 at 134-139; Rezek, *supra* note 65 at 307-315. Additionally, for reasons of military necessity or public safety, a State may restrict or prohibit other international aircraft from flying over certain areas of its territory, provided no distinction is made between such aircraft. See *Chicago Convention, ibid.*, Article 9.
69 See *Chicago Convention, ibid.*, Article 6. Greater freedom of movement is allowed by means of bilateral (or multilateral) agreements between States. Annexed to the Convention, the International Air Services Transit Agreement and the International Air Transport Agreement divide the "freedom of the air" into five categories. For more on the subject, see Cheng, *supra* note 48, ch. 5 at 8-17.
70 See *Chicago Convention, ibid.*, Article 11.
71 See *Charter, supra* note 48 at para.3 (a).
72 States have not accepted any obligation to provide such services beyond their sovereign territory, but on the basis of Regional Air Navigation Plans might accept to do so.
73 See M.Bartkowiski, "Responsibility for Air Navigation (ATM) in Europe" (1996) XXI:I *Annals of Air and Space Law* 45 at 47.
74 See F.Schubert, "Réflexions sur la Responsabilité dans le Cadre du GNSS" (1997) 45:180 *Revue Navigation* 417 at 417-418.
75 Kotaite, *supra* note 55 at 201. See also Rattray, *supra* note 12 at 4. But see Milde, *supra* note 61 at 211.
76 See ICAO, *Report of the First Meeting of the Secretariat Study Group on Legal Aspects of CNS/ATM Systems*, ICAO SSG-CNS/I-Report (9 April 1999) at para.3.8.4 [hereinafter *Study Group I Report*].
77 *Charter,* supra note 48 at para.2.
78 See *Study Group I Report, supra* note 76 at para.3.8.3; Milde, *supra* note 61 at 201.
79 See M.Milde, "Legal Aspects of Future Air Navigation Systems" (1987) XII *Annals of Air and Space Law* 87 at 92.
80 Milde, *supra* note 61 at 198.

81 See *Global Plan, supra* note 50, vol. 1 at para.11.2.5. See J.A.Mendez, "Cuestiones Técnicas y Jurídicas sobre los Nuevos Sistemas de Comunicaciones en la Navegación Aérea" in *La Aviación Civil Internacional y el Derecho Aeronáutico Hacia el Siglo XXI* (Buenos Aires: ALADA, 1994) 161 at 166, whereby the author expresses the need to reach a compromise between the relativism of the principle of sovereignty and the common benefit of mankind, so that the implementation of the CNS/ATM systems does not turn into a means of "subjugation" of developing nations.

82 See *Huang, supra* note 3 at 590.

83 See *Global Plan, supra* note 50, vol. 1 at para.11.2.7.4.

84 WW/IMP, "Assistance Requirements of States for CNS/ATM Implementation", ICAO WW/IMP-WP/27 (11 May 1998).

85 See *WW/IMP Report, supra* note 18 at para.5.1.1.

86 See *ibid.*, Conclusion 4/2 (a), (b).

87 See ICAO, *Statement of ICAO Policy on CNS/ATM Systems Implementation and Operation,* ICAO Doc. LC/29 - WP/3-2 (28 March 1994) at para.3. See also Kotaite, *supra* note 55 at 197-199.

88 For more information on ICAO's role under Chapter XV, see below, Chapter 7, Administration at 144ff.

89 See Kotaite, *supra* note 55 at 197; Huang, *supra* note 3 at 591.

90 See *Charter, supra* note 48 at para.5 (2).

91 See *Global Plan, supra* note 50, vol. 1 at para.11.2.6.2.

7 Other Legal Issues

Certification

Closely related with the principle of compatibility of GNSS with the Chicago Convention is the issue of certification. Like all air navigation facilities, GNSS requires certification by the relevant authorities to ensure compliance with navigation performance requirements related to the safety of international civil aviation.[1] Therefore an adequate system of ICAO SARPs on GNSS should not only cover the performance criteria of avionics and ground facilities, training and licensing requirements, but also satellite components and signal-in-space, as well as the system as a whole.[2]

During the discussions in the LTEP, views were expressed concerning the desirability of creating certain minimum standards from which no derogation would be possible, by application of Article 12 of the Convention. Said article vests the ICAO Council with binding powers to lay down regulations over the high seas. It was inferred that in the regulation of GNSS such powers could be, by analogy, extended to outer space, which is similarly not subject to claims of sovereignty.[3]

Views to the contrary compared the issue to an "unfounded mandate",[4] since the applicability without exception of the rules of the air contained in Annex 2 to the airspace over the high seas has an explicit constitutional basis in the Chicago Convention "and no constitutional basis exists for other purposes".[5]

The matter was finally settled by the Panel having determined that certification would take place in accordance with ICAO standards which, if not met, would allow for the application of Article 33, whereby contracting States may decline to recognise the validity of a certificate which does not comply with the minimum standards.[6] One commentator has challenged the application of said provision arguing that it has no relevance to the signal-in-space provider, but specifically refers to the recognition of certificates of airworthiness and competency, and licenses. Moreover, it does not impose a duty to reject such certificates and licences if such standards are not met,[7] but merely obliges States to recognise them if the minimum ICAO standards are complied with.

With a view to ensuring high integrity of GNSS related SARPs and limiting the number of differences filed, it has been recommended by the LTEP that signal-in-space provider States and provider international organisations be involved in the ICAO verification and validation process.[8]

Following a proposal in the Rapporteur's Report to the 29th Session of the Legal Committee, the possibility of ICAO playing an active role in the certification process was considered by the Panel.[9] Nevertheless, a majority was of the opinion that certification should fall on the sovereign States, being not the current practice of ICAO to certify equipment or services.[10] Upon further debate, it was agreed that ICAO could have a role in providing a forum for the exchange of information on GNSS certification.[11] A recommendation was adopted accordingly.[12]

The LTEP has further recommended that the State of Registry should continue to ensure that GNSS avionics, ground facilities, training and licensing requirements comply with ICAO SARPs. States providing signals-in-space, or under whose jurisdiction such signals are provided,[13] shall certify the signal-in-space by attesting it is in conformity with SARPs.[14]

Moreover, each State should define and ensure the application of safety regulations for the use of the signal-in-space as part of air traffic services in its sovereign airspace. When authorising GNSS-based air navigation services in its airspace, States need providers to demonstrate compliance of the elements with ICAO SARPs. Any additional information which may be required for this purpose should be made available through ICAO. Other sources, including bilateral or multilateral arrangements, and NOTAMs[15] may be used in addition thereto.[16]

Liability

The most complex of all the legal challenges raised by GNSS, as evidenced by the numerous debates which have dealt with the issue in various international fora, liability is in fact the main drive of the controversy pertaining to the need for an international convention as a long-term solution for the GNSS legal framework. An examination of the different opinions expressed so far discloses two distinct views:

On the one hand, there are those who believe it is premature to attempt to devise a specific liability regime for GNSS[17] which should take into account the practical experience in the commercialisation of the services as they develop.[18] In support of this view stand particularly those States operating the space segment for GNSS, as well as the few who already make use of such signals for air navigation purposes in their territories. Their arguments are presented as follows:

a) there are no differences between GNSS and other air navigation aids;[19]
b) no new legal problem has been identified yet to justify the development of supplementary provisions, considering that there is nothing inherent in CNS/ATM systems which is inconsistent with the Chicago Convention, and that there is a general agreement that there is no legal obstacle to their implementation;[20]
c) the current ICAO SARPs system is adequate and sufficient;[21]
d) the related subject of ATC liability has been on the agenda of the Legal Committee for more than three decades and there is no indication of an overriding necessity for an international framework;[22]
e) different liability regimes presently exist to cover the individual liabilities of each of the numerous players involved in the provision and operation of the GNSS,[23] such as the Warsaw Convention[24] or the Rome Convention;[25]
f) the current domestic legislation already provides an acceptable liability regime;[26]
g) States have gradually renounced the defence of sovereign immunity under specified conditions and can be held liable for damages in the same manner as private individuals;[27]
h) most victims of aviation accidents can obtain compensation from the air carrier, therefore there is no need for them to engage in multiple and complex actions in several jurisdictions;[28]
i) the issue of liability is in reality a matter of insurance cost, practical commercial experience having indicated that premiums are actually reduced when the service provider uses GNSS;[29]

j) the technology relating to the long-term GNSS is still evolving, and its characteristics and elements are far from clear at present. Only when there is a clear conception of what may constitute the long-term GNSS, will it be possible to say if additional law is needed.[30]

On the other hand, strong arguments have been put forward by those who believe there is a need for the adoption of a new international convention which would, in a simple, clear and straightforward manner, allow for the proper allocation of liabilities between the different partners involved in the provision and operation of the GNSS. Accordingly:

a) GNSS is indeed different from conventional terrestrial navigation aids in the sense that the total system will no longer be under the control of a particular State which undertook to provide air navigation services in its sovereign territory;[31]

b) the multilateral nature of the system and the internationalisation of its elements increase the complexity of related actions and the likelihood of procedural law problems. "Several layers of interconnected liabilities, unavoidable multiple, parallel, successive and recourse claims in various substantial legal regimes and in different countries, ... likely to extend endlessly into time"[32] are expected;

c) due to the multiplicity of actors and the range of defences available, a risk exists that victims of an accident involving failure or malfunction of the GNSS services will not be able to obtain full compensation when bringing action. Therefore, a victim-oriented approach in line with modern standards should be adopted with a view to ensuring prompt, adequate and effective compensation;[33]

d) SARPs may provide technical assurances as regards accuracy, integrity, availability and continuity for systems which are certified but cannot address broader institutional and liability issues, thus the need for another legal tool, such as an international convention, to regulate the relationship between providers and users of the signal-in-space;[34]

e) the application of the doctrine of sovereign immunity may render court action against a foreign State or foreign governmental entities providing ATC or GNSS signals, facilities and services in countries other than their home States difficult or impossible in the sense that they may lawfully refuse to submit to the jurisdiction of the court;[35]

f) Since technology relating to the long-term GNSS is evolving, it is necessary to agree on basic assumptions regarding the system's characteristics in order to discuss the issue of liability.[36]

Due to the dichotomy of views, the following concepts have been recommended by the LTEP to be further studied: i) fair, prompt and adequate compensation; ii) disclaimer of liability; iii) sovereign immunity from jurisdiction; iv) physical damage, economic loss and mental injury; v) joint and several liability; vi) recourse action mechanism; vii) channelling of liability; viii) creation of an international fund (as an additional possibility or an option); and ix) the two-tier concept, namely strict liability up to a limit to be defined and fault liability above the ceiling without numerical limits.[37]

As an introduction to that study, the implications of Article 28 of the Chicago Convention have been usefully examined by the ICAO Secretariat Study Group on the Legal Aspects of CNS/ATM Systems, during its first and second meetings held in April and October, 1999.

For purposes of these considerations, a distinction should be drawn between the responsibility of a State to provide air navigation services as an obligation or commitment under public international law, and the liability arising out of the breach of such an obligation, causing damage for which the State is actionable, according to the domestic rules which apply. Accordingly, Article 28 only regulates the relationship between sovereign States. Hence it does not give a cause of action to private parties, whose claims for compensation in case of damages must be handled by the applicable domestic law.[38]

The Implications of Article 28

As stated in the preceding pages, under Article 28 (a) of the Chicago Convention, each contracting State has undertaken, so far as it may find practicable, to provide in its territory air navigation facilities in accordance with ICAO SARPs. However, the mechanisms by which a State may fulfil this obligation have not been prescribed by the Convention.[39] Hence, in the exercise of their sovereignty, States may choose to delegate, in total or in part, the technical provision of air navigation services to a third party.[40]

At the national level, some States have delegated the responsibility of service provision to private entities or autonomous authorities. At the international level, precedents also exist of States entering into arrangements, for reasons of efficiency and economic benefit, which delegate to other States and/or joint agencies or international operating structures, such as ASECNA, COCESNA[41] or Eurocontrol, the provision of air navigation services within their sovereign airspace.[42] It has also been proclaimed to be perfectly proper that a State uses the services of a foreign provider of signals-in-space for providing air navigation services in its airspace.[43]

It should be duly noted, however, that regardless of the organisational structure under which such services are provided, the sovereign State remains ultimately responsible for their provision in its airspace,[44] namely for setting and maintaining the standards and for the quality of the services provided.[45]

In principle, therefore, the responsibilities of States with respect to satellite-based air navigation services refer to the entire system, including signals-in-space, augmentation services and other components. A question has been raised, however, whether the implementation of GNSS would represent any fundamental differences thereto, since with the introduction of GNSS, the core elements of the system, in particular the space segment, will no longer be under the territorial control of these States which undertook responsibility under Article 28.[46]

Some have held the view that the responsibility delegated to the provider of the signal-in-space is "purely functional, in the sense that it is limited to the technical provision of a given service",[47] the delegating (Article 28) State retaining the ultimate responsibility. In this regard, as in most cases where ATC agencies have been corporatised, it remains solely and directly liable for damages caused by the negligence of the provider State, even in the absence of fault or negligence of its own, although maintaining the right to recover the value of the compensation paid, if possible.[48]

A view to the contrary points out that the responsibilities of the Article 28 State are essentially of a regulatory and supervisory nature.[49] Before authorising the use of the signal-in-space in its territory, States must satisfy themselves that they comply with ICAO SARPs. They should also take appropriate measures to monitor signal compliance with SARPs on a permanent basis[50] and must issue proper warnings in case of disruption. Therefore they are capable of being held liable only if demonstrated they have failed to carry out these responsibilities or have not acted with reasonable care. A recommendation has been adopted by the LTEP, which provides that "signals should be recorded for purposes of evidence in accordance with ICAO SARPs".[51]

Finally, concerns have been voiced as to the situation where a State has expressly forbidden the use of the signal-in-space in its territory for navigation purposes, when it appears legitimate that it should bear no liability in the case of a GNSS failure related accident.[52] Accordingly, liability rules will apply if the State has approved, through a regulatory act, the use of GNSS signals in its airspace, which are then considered to form part of its own air navigation infrastructure.

In a pragmatic point of view, with reference to the doctrine of sovereign immunity, which may constitute an obstacle to bringing the provider of the signal-in-space into legal proceedings before the court where the victim has brought action, Roderick van Dam has stated that:

> If the interpretation of Article 28 followed is the one that supports that a non-provider State would not retain liability if it has performed its duties correctly, the victims may be at a loss since any recourses against the primary signal provider are most likely to be unsuccessful (unless the victim is a U.S. citizen). If the notion of the ultimate liability is retained, it would then be the non-providing States who would bite the liability bullet without any serious chance of successful recourses against the primary signal provider.[53]

In summary, those who believe that Article 28 provides sufficient tools for the implementation of the systems claim that, as long as the signals-in-space are provided in accordance with ICAO SARPs, and redundancy in air navigation facilities exists as a practical remedy in case of malfunctioning, no additional arrangements are necessary between providers and users.[54] The view appears to be oblivious of the extreme financial burden of implementing and maintaining two parallel systems.

On the other hand, there are those who believe that the current legal system has not adequately regulated the matter, due to "the lack of certitude regarding the interpretation of Article 28 itself".[55] Taking into consideration that certification provides only technical assurances and cannot address broader liability issues, various possible solutions have been put forward to supplement the lacunae of Article 28. Whereas an international convention providing for the proper allocation of liabilities between multiple actors might be the ideal solution, an amendment of the Chicago Convention is clearly not the way forward.[56] Nevertheless, additional legal arrangements whereby a link is established between the provider of signals-in-space and the user State, with the appropriate delegation of duties, are unquestionably necessary to deal with the disparity between responsibility and loss of control, and allow for the proper allocation of liabilities.[57]

Reviewing the discussion in the Secretariat Study Group, it is appears that the extent of the responsibilities undertaken by States under Article 28 still remains unclear. A conclusion had already been reached at the first meeting that the implementation of GNSS leaves unaffected the responsibility of States under Article 28 of the Convention.[58] However, upon consideration of the controversial issue of control, the Group agreed to further examine the implications of said article for States, under national and international law, when authorising GNSS for use in their airspace.[59] Particularly, "whether and to what extent there may be a partial or total 'release of liability' when 'outsourcing' the provision of GNSS signals, facilities and services to a foreign entity". And finally, "to what extent international rules may be required to deal with the interaction of the parties".[60]

The Current Liability Regime

With a view to determining to what extent the current liability regime would be adequate to deal with the implementation of GNSS, the Study Group invited its members "to inform the Secretariat of the legal rules in their respective jurisdictions applicable to claims against ATC, and which would likely be applicable in case of malfunctioning or failure of satellite signals used for navigation purposes".[61]

The applicable national law in Australia,[62] Canada,[63] France,[64] Italy,[65] the United Kingdom[66] and the U.S.[67] was reviewed to show that it is in these cases based on fault principles. Whereas ATC liability is governed by the law of torts in the common law jurisdictions, the concept of non-contractual responsibility will apply in the civil law jurisdictions. With the exception of France, where *faute lourde* or gross negligence is a requirement, it is particularly based on negligence (wrongful act or omission), and requires proof of fault of the ATC agency, or its employees or agents.[68] State liability is unlimited and victims could be compensated in full. The defences of contributory negligence, third party's fault and *force majeure* are usually available. Finally, personal negligence on the part of air traffic controllers could give rise to civil and criminal liability in France.

Although the reviewed legal systems were considered to be reasonably adequate and could provide satisfactory solutions to the liability issues arising from a GNSS failure related accident, the Group concluded that "the procedural rules and, in particular, the applicable rules on jurisdiction may not be adequate to bring all parties to the court in order to ensure prompt and equitable compensation in these cases".[69] In the words of David J.A. Stoplar, Assistant Legal Advisor at the CAA/NATS Ltd.:

> The victim would have to begin actions in a number of different jurisdictions, different legal regimes would apply, and no court would have all the relevant actors before it ... The litigation would be at best prolonged and expensive: at worst the victim could be denied redress because the truth of the matter was obfuscated in the course of the complex legal process.[70]

However, during the preceding debates, a couple of experts insisted on that the multiplicity of potential defendants does not present a problem in the jurisdiction of their respective States, due to the existence of rules of procedure "which allow for the joiner of claims and parties and the consolidation of proceedings, thus reducing the apparent complexity of litigation".[71] Particularly, M.Jennison pointed out that:

> Tort law provides rules for determining the legal relationships of tortfeasors, and judges and juries have shown that they can sort out even complicated, extensive, and intricate relationships and chains of events on the basis of evidence presented by the parties.[72]

The Study Group specifically recognised that additional problems may arise as a consequence of the application of the doctrine of sovereign immunity and related principles, whereby States may refuse to appear before the court seized of the case in a foreign jurisdiction. The issue gains particular relevance when it comes to the determination of the liability of the signal-in-space provider, currently exclusively submitted to its own domestic law. Accordingly, the liability of the U.S. government under U.S. law and statutes merits special consideration.

Liability of the U.S. Government Under U.S. Law – The Federal Tort Claims Act

Under the Federal Tort Claims Act (FTCA),[73] the federal government has waved, with certain exceptions, its sovereign immunity from liability in tort for the acts of its employees acting within the scope of their office. Accordingly:

> The district courts ... shall have exclusive jurisdiction of civil actions on claims against the U.S., for money damages ..., for injury or loss of property, or personal injury or death caused by the negligent or wrongful act or omission of any employee of the Government while acting within the scope of his office or employment, under circumstances where the U.S., if a private person, would be liable to the claimant in accordance with the law of the place where the act or omission occurred.[74]

An early decision of the Supreme Court in *Indian Towing Co. v. U.S.*,[75] whereby the Coast Guard was held liable for damages resulting from the negligent operation of a lighthouse, led to the application of liability for negligence in air traffic control operations.[76] Thus, to prevail upon a claim of negligence of the air traffic controller, the plaintiff must establish the following elements: i) duty of reasonable care; ii) breach of that duty; iii) and proximate damages resulting from that breach.[77] Contributory and comparative negligence defences are available, and depending on the applicable state law, will allow a court to weigh the respective fault of the parties and assign liability accordingly.[78]

The Act did not waive the sovereign immunity of the U.S. in all respects. Indeed, it includes a number of exceptions, all of which might have an impact upon GNSS related liability claims.

Discretionary Function Exception

The following provision of the FTCA exempts from statutory liability:

> Any claim based upon an act or omission of an employee of the Government, exercising due care, in the execution of a statute or regulation, whether or not such statute or regulation be valid, or based upon *the exercise or performance or the failure to exercise or perform a discretionary function or duty* on the part of a federal agency or an employee of the Government, whether or not the discretion involved be abused.[79]

Courts have interpreted the discretionary function exception on the basis of a simple but deceptive distinction, namely that there is policy level discretion available to the individual at the planning stage, but there is no such discretion at the implementation or operational level.[80]

The evolution of case law has established that the exception "insulates from liability only those governmental actions and decisions that *involve an element of judgement or choice* and that are *based on public policy considerations*".[81] Thus, the exception will not apply "when a federal statute, regulation or policy specifically prescribes a course of action for the employee to follow", and he "has no rightful option but to adhere to the directive".[82]

In view of the current interpretation, it has been submitted that, in principle, the U.S. government would be held liable for tort damages if it were established that the failure or malfunctioning of the GPS signal was the proximate cause of an accident. However, the decision to provide such a signal, or to provide it at a particular accuracy level would be construed as discretionary[83] and therefore be protected. Others have maintained that once the decision is made and the provision of the signal-in-space continues, the maintenance of the appropriate standards would be considered "operational".[84]

Foreign Country Exception

American courts have dismissed claims for lack of subject-matter jurisdiction, holding them to be barred by the FTCA's foreign country exception, which states that the statute's waiver of sovereign immunity does not apply to "[a]ny claim arising in a foreign country".[85]

Views have been expressed that where a claim arises is not always the scene of the accident.[86] Hence, in a claim related to GPS, it is the place where the wrongful act or negligent conduct took place which is to be taken into account. In the words of M. Jennison:

> Claimants would be expected to argue that the negligence occurred in the manufacture of the satellites in a U.S. plant, in their launch at a U.S. range, their control and monitoring at the Air Force operations centre at Colorado Springs, or perhaps *even in outer space, which is not subject to any State's sovereignty.*[87]

In a correlative ATC situation, namely the *In re Paris Air Crash of March 31, 1974* case, the court indeed held that the fact that the accident occurred in France did not bar suit in the U.S., because the wrongful act was alleged to be the approval of a Certificate of Inspection in California.[88]

Nevertheless, in the recent *Smith v. U.S.* case, the Supreme Court held that "the FTCA does not apply to tortious acts or omissions occurring in the sovereignless region of Antarctica", and that "the ordinary meaning of 'foreign country' includes Antarctica, even though it has no recognised government".[89] This interpretation was based on the language and structure of the statute itself, and on a presumption against the extraterritorial application of U.S. statutes, whereby it is assumed that courts are prohibited from extending or narrowing the waivers of sovereign immunity beyond what Congress intended.[90]

Justice J. Stevens filed a dissenting opinion, in which he points out that "Antarctica is just one of three vast sovereignless places where the negligence of federal agents may cause death or physical injury". He specifically makes parallel to outer space, a region "far beyond the jurisdictional boundaries which were familiar to the Congress that enacted the FTCA in 1946".[91] In his view, the presumption against the extraterritorial application federal statutes has no bearing on the case:

> The fact that Congress intended and understood the broad language of those provisions to extend beyond the territory of the U.S. is demonstrated by its enactment of two express exceptions. One of those is ... the foreign country exclusion in 2680 (k). The other is the exclusion in 2680 (d) for claims asserted under the Suits in Admiralty Act or the Public Vessels Act. Without that exclusion, a party with a claim against the U.S. cognisable under either of those venerable statutes would have had the right to elect the pre-existing remedy or the newly enacted FTCA remedy. Quite obviously, that exclusion would have been unnecessary if the FTCA waiver did not extend to the sovereignless expanses of the high seas.[92]

In brief, the reasoning of the Court seems more consistent with the narrow interpretation that the FTCA has an "exclusive domestic focus" and that it applies "only within the territorial jurisdiction of the U.S.".[93]

Consequently, as far as GNSS is concerned, it has been submitted that according to the current interpretation of the Supreme Court, outer space would also be included in the meaning of the expression "foreign country", as defined by the FTCA. Government immunity would probably prevail in claims arising out of a failure or malfunctioning of the GPS signal-in-space.[94]

Combatant Activity Exception

The third exception to the FTCA waiver of sovereign immunity is related to the activities of the U.S. Armed Forces in time of war.[95] Concerns have especially been raised as to the possibility of the GPS being shut down in a national emergency, when the U.S. would be completely shielded from liability.

As a last remedy, the Foreign Claims Act[96] and the Military Claims Act[97] could provide an administrative means of recovery to inhabitants of foreign countries and U.S. citizens who may file claims against the U.S. for property damage, injury or death caused by non-combatant activities or by members and/or civilian employees of the Armed Forces acting in an official capacity.

Conclusion

From all the above-mentioned, it rests clear that suing the U.S. government in U.S. courts for damages arising out of GPS related activities will not be an easy or promising task. There is a possibility that claims will be barred by the FTCA's exceptions and that sovereign immunity will prevail.

With no guarantee of the successful application of the waivers, the absence of any legal instrument addressing the liability of the signal-in-space provider gives rise to the greatest concern of the international community. In the time consuming and expensive multiplicity of successive, parallel and recourse actions, Article 28 States fear that "the liability wheel could stop running at their doorstep".[98]

Hence, an adequate recourse action mechanism is called for. An international convention under which liability issues could be resolved in a simple and speedy procedure is the appropriate long-term solution. Meanwhile, specific arrangements with the U.S. are necessary. An analogy could therefore be made with the delegation of air traffic services where, in principle, the State or entity performing the services recognises liability when negligent.[99] As previously stated, the concept of channelling of liability might be useful in the sense that it would call for contractual agreements between the various components of the system, where individual performance criteria would be established, allowing for the extent of liability to be easily identified in the context of a contractual framework.[100]

Furthermore, in most States, courts will, in principle, recognise the immunity of a foreign State from jurisdiction, and will not be able to enforce any claims, since the property of that State is not subject to execution. Exceptions do exist, and are usually related to commercial transactions undertaken by the State.[101] To take just one example, this seems to be the case in Canada where, apart from the commercial activity exception provided under Section 5 of the State Immunity Act, Section 6 provides that a foreign State is not immune from jurisdiction of a court in any proceedings that relate to any death or personal injury, or any damage to or loss of property that occurs in Canada.[102]

In a broad context, therefore, irrespective of the organisational structure of the service provider, it should be ensured that any foreign State, group of foreign States or foreign governmental entities providing ATC or GNSS signals, facilities and services remain accountable for their actions and omissions. Particularly, in case of fault or negligence of the State or its agents in the provision of signals-in-space, services or facilities, the doctrine of sovereign immunity must not constitute an obstacle to bringing all parties into legal proceedings before the court where the victim has brought action. Hence, adequate remedies to obtain prompt, just and adequate compensation must be provided by States in their national laws.[103]

Other Existing Compensation Channels

With the introduction of the GNSS, the legal complexities which may arise in the event of an accident are profoundly exacerbated by the multiplicity of actors involved. Several layers of interconnected liabilities can be expected to further complicate and extend legal proceedings, and victims might need to engage in multiple parallel and consecutive legal actions to attempt recovery of the full value of the damage.[104]

Bringing all potential defendants into a single action might be a solution frustrated in the mere intent for most States would not be willing to submit to a foreign jurisdiction nor will ever comply with a judgement pronounced by such court. Moreover, some legal systems also make a distinction between private and public entities for reasons of establishing court jurisdiction.[105]

Commenting on the issue, Dr. Francis Schubert, Corporate Secretary for Swisscontrol, has asserted that:

> [L]egal proceedings will also be complicated by the fact that there will normally be more than one victim and the different claimants may elect different

compensation channels or seek compensation from the same defendant, but in different countries ... In most situations, it will not be possible to settle the final allocation of liabilities through direct actions alone. Some of the defendants may be compelled to compensate for damages while no negligence can be blamed upon them or while other parties may be partly or totally responsible for the accident. This will unavoidably lead to recourse actions ... , the objective of [which is] to recover a part or the totality of the amount the initiator had to pay itself to passengers, third parties on the surface or the air carrier in first instance.[106]

In the scenario so clearly described above, possible defendants include, *inter alia*:

a) the signal-in-space provider (State, group of States or International Organisation);
b) the augmentation provider (State, group of States or International Organisation);
c) the Article 28 State having certified the GNSS equipment and authorised the use of GNSS in its airspace;
d) the ATC agency;
e) the air carrier;
f) the aircraft operator;
g) the State of registry of the aircraft;
h) the equipment and the components manufacturers;[107]
i) third parties interfering with the signal; and
j) the pilot-in-command.

Although a variety of compensation channels exists and may be considered reasonably adequate to address all possible legal complexities, the lack of uniformity in the multiplicity of individual legal regimes, national or international, which might be applicable to different actors in different jurisdictions may result in "uncontrollable conflicts of law and jurisdiction, an endless succession of legal proceedings, and, possibly, partial or total denial of justice".[108]

As seen, the current liability regime which would be applicable to GNSS claims is governed by the domestic law of the State concerned. Other compensation channels also exist. Consideration will be given here to other applicable international law.

The Warsaw Convention and the Montreal Convention

In a GNSS environment, a situation could arise where the air carrier could be held liable, for example, for damages arising out of the use of a faulty signal, despite warnings, or an unauthorised signal.[109]

In this regard, an action against the air carrier engaged in international transportation under the Warsaw Convention appears to be the easiest channel available to the passenger at present. Subject to a regime of presumed fault with a reversed burden of proof, the air carrier is liable for damage sustained in the event of death or any other bodily injury suffered by a passenger, unless he proves that he and his agents have taken all necessary measures to avoid the damage or that it was impossible for him or them to take such measures.[110]

Article 22 places a ceiling on the liability of the air carrier in all suits covered by the Convention. Unrealistically low, they impose the greatest risk of incomplete compensation. However, recovery will not be limited to the amounts stated in the Convention, if the plaintiff can prove that damages were caused by the wilful misconduct of the air carrier.[111]

The urgent need to modernise and consolidate the Warsaw Convention and related instruments led to the development of the Montreal Convention,[112] opened for signature on 28 May, 1999. The new instrument will introduce a two-tier liability regime, namely, strict liability irrespective of the carrier's fault up to 100,000 Special Drawing Rights (SDR), and unlimited liability above that limit. The carrier shall not be liable for damages exceeding 100,000 SDR if he proves that: i) such damage was not due to the negligence or other wrongful act or omission of the carrier or its servants or agents; or ii) such damage was solely due to the negligence or other wrongful act or omission of a third party.[113]

At the option of the plaintiff, an action for damages under the Warsaw Convention must be brought in the territory of a contracting Party, either before the Court having jurisdiction where the carrier is ordinarily resident, or has his principal place of business, or has an establishment by which the contract has been made or before the court having jurisdiction at the place of destination.[114] A fifth jurisdiction based on the "principal and permanent residence" of the passenger will also be available under the Montreal Convention.[115]

In the case of aircraft accidents resulting in death or injury of a passenger, the carrier shall, if required by its national law, make advance payments without delay to the person who is entitled to claim compensation in order to meet the immediate economic needs of such person. However, such advance payments shall not constitute a recognition of liability and may be offset against any amounts subsequently paid as damages by the carrier.[116]

Another innovation to be introduced by the new instrument is compulsory insurance. Accordingly, States shall require their carriers to maintain insurance covering their liability under the Convention.[117] Evidence thereof may also be required by any State Party into which the carrier operates.

The Montreal Convention will enter into force on the sixtieth day following the deposit of the thirtieth instrument of ratification, acceptance, approval or accession with the Depositary.[118]

The Rome Convention

Third parties on the surface may have a cause of action against the aircraft operator under the Rome Convention, upon proof only that the damage was caused by an aircraft in flight or by any person or thing falling therefrom.[119]

A strict liability regime therefore applies, but the aircraft operator is entitled to the defence of contributory negligence and will be exonerated from liability to the extent that he proves that the negligence or other wrongful act or omission of the person who suffered damage, or of the latter's servants or agents, contributed to the damage.[120] No limitation of liability shall apply, however, if the claimant proves that the damage was caused by a deliberate action or omission of the operator.[121]

It has been submitted that the Convention might not have much relevance to claims arising out of a GNSS related accident, because of the relatively low number of ratifications to the instrument, which requires that both the State of Registry of the concerned aircraft and the State over the territory of which the accident occurs are parties to the Convention.[122] Furthermore, as previously stated, in a regime of strict liability the cause of the accident need not be demonstrated by the claimant, unless there is an interest in breaking the limits of liability. In this case, the claimant would have to prove, for example, that a warning of system failure had deliberately been ignored by the aircrew, or else that they had other means to detect the proximity of the other aircraft.[123]

The Liability Convention

Different interpretations of the Liability Convention[124] exist as to whether or not it provides a basis for claims against the signal-in-space provider for damages arising out of a GNSS failure or malfunction.

The Convention provides that a launching State shall be absolutely liable for damage caused by its space object on the surface of Earth or to aircraft in flight. The term "space object", as defined in Article I, includes component parts as well as its launch vehicle and parts thereof. Whether a signal emitted therefrom is to be considered a space object has been subject to much debate.

The predominant view, which receives our support, is that the Convention aims to cover only direct physical impact with a space object. Indirect or consequential damage such as faulty transmission or reception of a signal generated by a space object would not be recoverable under the Convention.[125]

Regardless of the interpretation, an important reminder is that claims under the Convention shall be made to a launching State through diplomatic channels. Any person, natural or juridical, would therefore file with the State of nationality who would present the claim on his or her behalf, and then await the diplomatic process before receiving any compensation for the damage sustained.[126]

International Fund for Compensation

The introduction of compulsory insurance and the related subject of the establishment of an international compensation fund have also been considered with respect to liability for GNSS services.[127] Accordingly, a recommendation was adopted by the LTEP which provides that in studies on the liability regime for GNSS, it should be taken into consideration that appropriate methods of risk coverage should be utilised so as to prevent frustration of legitimate claims.[128]

As far the air carrier is concerned, the Montreal Convention, not yet in force, is the first international air law instrument ever to make direct provision for compulsory insurance. As previously mentioned, States shall require their national carriers to maintain adequate insurance, covering their liability under the Convention. It should be duly noted that domestic legislation in a number of States already makes provision thereof.[129]

Some international operating agencies, such as ASECNA, COCESNA and Eurocontrol have also had recourse to insurance to cover their liability for damages sustained by users in the provision of air traffic services. A specific provision in this regard is contained in their respective constitutional instruments.[130]

Likewise, the signal-in-space provider should possess adequate risk coverage. It has been submitted that a State which authorises the use of GNSS as part of its air navigation infrastructure should satisfy itself that the provider is sufficiently covered.[131]

In the event that insurance does not cover or is insufficient to satisfy the claims for compensation for damages sustained in relation to GNSS, or a limited liability regime is in place within the framework of an international convention, the establishment of an international compensation fund has been suggested by some experts in the LTEP.[132] The proposed fund is envisaged along the lines of the International Convention on the Establishment of an International Compensation Fund for Oil Pollution Damage.[133] The said instrument purports to elaborate a compensation and indemnification system supplementary to the International Convention on Civil Liability for Oil Pollution Damage,[134] with a view to ensuring that full compensation will be available to victims of oil pollution incidents, and that the ship-owners are at the same time given relief in respect of the additional financial burdens imposed on them by the said Convention.[135] For these purposes:

> [T]he Fund shall pay compensation to any person suffering pollution damage if such person has been unable to obtain full and adequate compensation for the damage under the terms of the Liability Convention,
>
> (a) because no liability for the damage arises under the Liability Convention;
> (b) because the owner liable for the damage under the Liability Convention is financially incapable of meeting his obligations in full and any financial security that may be provided ... does not cover or is insufficient to satisfy the claims for compensation for the damage ... after having taken all reasonable steps to pursue the legal remedies available to him;
> (c) because the damage exceeds the owner's liability under the Liability Convention as limited pursuant to Article V, paragraph 1, of that Convention or under the terms of any other international Convention in force or open for signature, ratification or accession at the date of this Convention ...[136]

The Fund, which has separate legal personality and can be a party in legal proceedings,[137] shall incur no obligation if it proves that the pollution damage resulted: i) from an act of war, hostilities, civil war or insurrection; ii) wholly or partially either from an act or omission done with intent to cause damage by the person who suffered the damage or from the negligence of that person. In any event, the Fund shall be exonerated to the extent that the ship owner may have been exonerated under the Liability Convention.[138]

The Fund also indemnifies ship-owners for a part of their strict liability under the 1969 Convention, provided, however, that the Fund shall incur no obligation where the pollution damage resulted from the wilful misconduct of the owner himself.[139]

Contributions to the Fund shall be made in respect of each Contracting State by any person who, in the calendar year before the entry into force of the Convention for that party, received quantities of oil exceeding 150,000 tons, such contributions to be calculated on a "per ton" basis, as determined by the Assembly of the Fund.[140]

The concepts behind the Convention merit further consideration and could be adapted to suit the peculiarities of the GNSS.

Disclaimer of Liability

An important question which arose in the discussions of the LTEP concerns whether the prevailing practice in space telecommunications of broad liability disclaimers for signal failure due to telecommunications breakdowns should be allowed in contracts regarding satellite-based air navigation.[141]

A particular example thereof might be Inmarsat, the liability of which finds itself curtailed by Article XII of its Operating Agreement. Accordingly:

> Neither the Organisation, nor any Signatory in its capacity as such, nor any officer or employee of any of them, nor any member of the board of directors of any Signatory, nor any representative to any organ of the Organisation acting in the performance of their functions, shall be liable to any Signatory or to the Organisation for loss or damage sustained by reason of any *unavailability, delay or faultiness of telecommunications services* provided or to be provided pursuant to the Convention or this Agreement.[142]

In consistency with Article 36 of the ITU Convention, which stipulates that Members accept no responsibility towards users of the international telecommunication services, particularly as regards claims for damages,[143] the Inmarsat Terms and Conditions for the Utilisation of the Space Segment by Navigation Land Earth Stations and Mobile Earth Stations also contain disclaimers of liability of Inmarsat for loss due to telecommunications breakdowns. Moreover, they expressly require Signatories to obtain a corresponding disclaimer in their contracts with the earth station operators and service providers for provision of the services, if consistent with national law.[144]

In claims against Inmarsat by contractors or third parties, if the Organisation is required by a binding decision rendered by a competent tribunal or as a result of a settlement agreed to or concurred in by the Council to indemnify, the Signatories shall, to the extent that the claim is not satisfied by indemnification, insurance or other financial arrangements, pay to the Organisation the amount unsatisfied on the claim in proportion to their respective investment shares.[145]

In a practical example, EGNOS' space segment will use the transponders on the Inmarsat-III satellites leased to France Télecom and Deutsche Telekom, thereby establishing a contractual relationship among them.

It has been stated that since services are provided by Inmarsat on a contractual basis, the channelling effect of the disclaimers would finally lead to the end user, namely the aircraft operator which, in turn, would pass on the financial burden to the international traveller.[146]

In view of the situation described, concerns were voiced that the widespread use of disclaimers might not instil the desired level of confidence in States, since the signal-in-space provider might not be willing to accept liability for loss of signals. Opinions were expressed that safety is of paramount importance in the provision and operation of GNSS services and must not be compromised. States must be able to rely on the accuracy, availability, continuity and reliability of the signal-in-space the use of which they have authorised in their airspace. On the other hand, consideration should be given to the commercial reality of the communication services so that, despite the possible existence of such clauses, the parties concerned would be expected to exercise an adequate degree of prudence and discernment when making use of the liability disclaimers.[147]

The issue is therefore still unresolved. A recommendation has been adopted by the LTEP in this regard:

> The vital role of the signal transmitted by navigation satellites for the safety of international civil aviation could raise the question whether disclaimers of liability would be appropriate in the case of navigation satellites, particularly in cases involving accidental death or injury.[148]

Channelling of Liability

In direct relevance to the above-mentioned, the concept of channelling of liability, previously described in this chapter[149] gains particular relevance. In the proposed series of contractual arrangements to be signed between all actors involved in the provision, operation and use of the GNSS services, each and every one will assume its share of responsibility against specific performance requirements described therein. A transfer of liability to other parties should not weaken the duty of care of each actor.

A recommendation by the LTEP on the issue was put to an indicative vote, and adopted by a majority.[150] Accordingly, "[the Council] should encourage the study of the concept of addressing liability through a chain of contracts between GNSS actors as an approach, in particular, at regional level". Moreover, "a model for future contractual arrangements should embody the work done by the Panel in applying the relevant recommendations".[151]

Regime of Liability

Different approaches to a unified liability regime for GNSS have been identified. Particularly, while some are in favour of an unlimited liability regime, regardless of fault, others see a limitation of liability as a quid-pro-quo for no-fault liability.[152]

It has also been suggested, and especially recommended by the LTEP to be further studied, that the new trend in private international air law, namely a two-tier concept which includes strict liability up to a certain monetary threshold, and fault liability above that ceiling without numerical limits, could be extended to the compensation of GNSS related damages.[153]

The proposed concept would certainly not present a problem as far as the air carrier and the aircraft operator are concerned, in view of the fact that a number of air carriers have already voluntarily submitted to such a regime under national law or under the IATA Intercarrier Agreement. Moreover, the Montreal Convention, recently opened for signature, also embodies the principle.

On the other hand, ATC services are currently governed by a fault-based unlimited liability regime, and damages are, in most cases, related to human error. Nevertheless, it has been asserted that for safety reasons, the dependency relationship that will develop with the GNSS infrastructure, along with the higher possibility of damages arising out of a technical failure rather than human error, would definitely justify the move to a strict liability regime.[154]

Finally, it has been submitted that a victim-oriented approach in line with more modern standards should be adopted, with a view to ensuring fair, prompt and adequate compensation. In this regard, physical damage, economic loss and mental injury should all be contemplated.[155]

Administration

It has been affirmed that the organisational structure under which CNS/ATM systems and air navigation services are operated is fundamental to their financial viability. Particularly, the magnitude of the investments required for the implementation of the systems has determined that it is not convenient, feasible or practicable for a State to implement such system for its own sole use.[156] In these circumstances, increased financial and operational autonomy at the national level, as well as the adoption of a co-operative and multinational approach, are deemed essential for States to be able to reap additional benefits from the cost-effective implementation of the systems.[157]

In this respect, different implementation options for the provision of air navigation services are available to States at the national level, namely a government department, an autonomous public sector organisation or a private sector organisation.[158] At a multilateral level, States may benefit from international co-operation by means of the establishment of international operating agencies, joint charges collection agencies, multinational facilities and services, and joint-financing type-arrangements.[159]

Any option chosen will have a direct impact on cost recovery schemes and funding of the systems.

National Level

a) Government Department
b) Autonomous Authority
c) Private Sector Organisation

Until now, most air navigation services have been provided by an organisation within the government, such as a civil aviation authority or a government department with similar responsibilities. In this context, these services constitute only one of the many functions which could be assigned to the organisation, including regulatory and licensing activities.[160]

Funded by the government, sometimes through general taxation, any capital expenditure therein not only is subject to the government's approval, but must also compete with many other claims for government funds.[161] Therefore, the difficult financial situation which the majority of these organisations has been experiencing is, in great measure, the result of the pressure to finance other high priority services in the States, instead of civil aviation.[162] In this sense, charges levied for the services provided are known to be used by many governments for general purposes other than defraying the costs of the facilities and services.

The need to improve financial results and efficiency in the provision of air navigation services has prompted a possible solution in the form of an autonomous public sector organisation, established with the specific function of operating such services. Also referred to as an autonomous authority, it constitutes an independent entity that is granted operational and financial freedom, but remains under the overall ownership of the government,[163] responsible for monitoring its performance.[164]

The organisation charges for the services provided and uses such revenue to fund operating expenses and to finance capital expenditure. Although the government would normally provide finance capital, access to the private capital market could be allowed on a limited basis.[165]

In view of the potential economic benefits to be derived from their managerial flexibility, increased efficiency and financial transparency, States have been particularly advised to consider the establishment of autonomous authorities, where traffic density would permit the generation of user charges to make such entities self-sustaining.[166]

A third alternative, usually seen as a means to lessen the burden of heavy capital investment from the State, is privatisation.[167] In this regard, a distinction should be drawn between the different terms employed to describe various situations involving changes in the ownership and control in the provision of air navigation services. Accordingly, whereas the word "privatisation" always connotes full ownership or at least majority ownership of facilities and services in the hands of the private sector, reference should be made to "private participation" or "private involvement" whenever majority ownership remains with the State. Another approach to the management of air navigation services and facilities, which involves the creation, under a statute, of a legal entity outside the government is referred to as "corporatisation".[168] So far, the closest example to the establishment of a private sector organisation has been the commercialisation of air navigation services in Canada, through the creation of NAV CANADA as a non-share capital corporation.[169]

As a direct result of the monopolistic nature of air navigation services, privatisation calls for a number of safeguards. Obligations such as freedom of access, non-discrimination between categories of users, strict compliance with aviation safety and security standards set by the government and international agreements must be observed. Particularly, the principles contained in the Chicago Convention, the Annexes, ICAO Regional Air Navigation Plans, and ICAO policies and statements shall continue to apply.[170]

Whatever the organisational format elected, it must be recalled that the ultimate responsibility for the quality of service provision remains with the State. Therefore, the State does not ever abdicate of its safety oversight and economic regulatory role, and must continuously ensure compliance with the established international standards applicable in its territory. A recommendation in this regard has been recently adopted by the Conference on the Economics of Airports and Air Navigation Services convened by ICAO in June, 2000.[171] Hence, it is imperative that an adequate regulatory framework and enforcement mechanism be put in place so as to ensure that these obligations are properly addressed.

International Level

a) International Operating Agencies
b) Joint Charges Collection Agencies
c) Multinational Facilities and Services
d) Joint Financing Arrangements.

It has been continuously stated that international co-operation may be, in most circumstances, the most cost-effective and only realistic approach to the implementation of the CNS/ATM systems. A recommendation in this regard has been adopted by the Rio Conference, stressing the need for States to adopt a co-operative, multinational approach in order to ensure seamless and interoperable systems at the regional and global levels. Particularly, proper co-ordination will help avoid duplication of efforts and proliferation of system elements so as to reduce costs and enhance safety, while increasing operational efficiency.[172]

Experience indicates that technical and operational constraints associated with the provision of air traffic services can be helped by the establishment of international operating agencies, such as ASECNA, COCESNA or Eurocontrol. As an international autonomous authority, the agency would be tasked with the provision of air navigation services, principally route facilities and services, within a defined area on behalf of two or more States. In addition, it would also be responsible for the operation of charges collection systems for the services provided.[173]

A second option would be the creation of Joint Charges Collection Agencies with a view to facilitating and minimising costs involved in the collection of route charges levied for air navigation services. The agency would collect such charges on behalf of all participating States, including those overflown, which would each receive its corresponding share of the revenue. The agency's costs should not be deducted from these shares but added to the charges levied on users on behalf of each State.[174]

Thirdly, and particularly significant in the context of CNS/ATM systems, is the possibility of creating, within an ICAO regional plan, a multinational facility or service. The main purpose of this organisation would be to service international air navigation in an airspace extending beyond the airspace serviced by a single State. Participation of States should be formalised in an agreement to ensure the fair and equitable sharing of all costs involved, as well as cost recovery through user charges. Such agreement could take the form of an international treaty or an administrative agreement, the latter being less time-consuming and allowing for more flexibility in case of any subsequent modification therein.[175]

Finally, the joint financing of air navigation services and facilities is a possibility contemplated by Chapter XV of the Chicago Convention. It particularly serves situations where it might be extremely costly for a State to provide facilities and services for which it has just a minimal need, being not unreasonable that other States affected participate in the financing thereof.[176]

In principle, Article 69 of the Convention stipulates that if the ICAO Council is of the opinion that the air navigation facilities of a contracting State are not reasonably adequate for the safe, regular, efficient, and economical operation of international civil aviation, it shall consult with the State directly concerned and the States affected, and make recommendations for the purpose of remedying the situation.[177] In these circumstances, a State may conclude arrangements with the Council, where it may elect to bear all costs involved. If it does not so elect, the Council may agree, at the request of the State, to provide for all or a portion of the costs.[178] Additionally, upon the request of a contracting State, the Council may agree to provide, man, maintain, and administer any air navigation facilities required in its territory for the operation of air services of other contracting States, and may specify just and reasonable charges for the use of the facilities provided.[179]

Chapter XV also contains provisions regarding the assessment of funds. Accordingly, "the Council shall assess the capital funds required for the purposes of this Chapter in previously agreed proportions over a reasonable period of time to the contracting States consenting thereto, whose airlines use the facilities".[180] The Convention also admits of the Council making current expenditures for the purpose of financing airports and air navigation facilities,[181] and makes allowance for the provision of technical assistance to States.[182]

In practice, due to the magnitude of the investments involved in the implementation of the CNS/ATM systems, and particularly the global nature of the services, it has been submitted that a joint-financing-type agreement could be extremely useful for the provision and operation of the systems' elements. Accordingly, it could be carried out either by a single State on behalf of the other participating States, contracted to a service provider or commercial operator, or undertaken by a group of States which would jointly operate and provide the facilities and services concerned.[183]

A successful example, which has been specially recommended by the LTEP to be used as a model for the GNSS,[184] is the Danish and Icelandic Joint Financing Agreements.[185] Established in the form of multilateral agreements, they purport to regulate the overall operation, administration and financing of the services provided on behalf of the international community engaged in North Atlantic flights.[186] These services comprise air traffic control, communications and meteorology. Financial responsibility is assumed by a group of 23 States, including the two provider States, all parties to the agreements.[187]

At first, in consideration of the special benefits to be derived from the services, Denmark and Iceland accepted to bear 5 per cent of the costs thereof. The other 95 per cent were to be shared between the contracting States in proportion to the aeronautical benefit derived therefrom, and calculated, on a yearly basis, by the number of complete crossings performed in that year by the contracting States' civil aircraft on routes between North America and Europe north of the 40th parallel North.[188] Such crossings were redefined in 1982.

User charges to be levied on all civil aircraft flying over the defined region were eventually introduced in 1974, at which time the United Kingdom agreed to act as an agent in the billing and collection of the charges.[189]

Currently, all civil aircraft flying across the North Atlantic north of 45° N latitude, whether or not their governments participate in the agreements, must pay a user charge for the services provided. Other costs not allocable to civil aviation are shared among the contracting parties.[190]

The responsibility for the administration of the agreements rests with the ICAO Council and the Secretary General. However, the participating States exercise full control through a Council Committee - the Joint-Support Committee – which advises in carrying out ICAO responsibilities under the agreements, ensures that the procedures established are followed, examines the financial and technical aspects of new requirements, and makes recommendations to the Council.[191]

In brief, the possibility of using the experience of ICAO, as a neutral organisation, to solve common difficulties in the administration of complicated air navigation services, along with the characteristic legal and structural flexibility of the arrangements, make the DNE/ICE agreements a transparent model of fairness and equity, with clearly defined needs and objectives, and therefore, an option particularly interesting in the context of the global navigation satellite system.

Cost Recovery

Whatever administrative mechanism is chosen by States at the national and multilateral levels for the provision of air navigation services, it is recommended that the costs of implementing and operating the CNS/ATM systems components be recovered through the medium of user charges[192] in conformity with basic ICAO airport and air navigation services cost recovery policy.[193]

Article 15 establishes the basic principles on the issue of cost recovery, namely:

a) uniform conditions shall apply to the use, by aircraft of all contracting States, of airport and air navigation facilities provided in the territory of a contracting State;
b) any charge imposed for the use of such airports and air navigation facilities by the aircraft of other contracting State shall not be higher than those that would be paid by its national aircraft engaged in similar operations;
c) no fees, dues or other charges shall be imposed by any contracting State in respect solely of the right of transit over or entry into or exit from its territory of any aircraft of a contracting State.

The principles set forth in the ICAO's Policies on Charges for Airports and Air Navigation Services[194] constitute valuable guidance on general and specific aspects of cost recovery, and could be summarised as follows:

a) providers of air navigation services for international use may require the users to pay their share of the related costs. However, international civil aviation should not be asked to meet costs which are not properly allocable to it.[195]
b) payment may still be required from users when utilisation of the services provided does not take place over the territory of the provider State;[196]
c) the cost to be shared is the full cost of providing the services, "including appropriate amounts for cost of capital and depreciation of assets, as well as the costs of maintenance, operation, management and administration".[197]

In fact, in response to the principle of cost-relatedness for charges, and in order to protect users from being charged for services or facilities which are not yet operational, it is recommended that user charges be applied only against defraying costs of existing services and facilities, including satellite services, implemented under the Regional Air Navigation Plan.[198]

Nevertheless, a recent trend has been observed in many States in respect to the pre-funding of future facilities and services requiring major capital investments, such as the implementation of the CNS/ATM systems. Accordingly, the ICAO Council considers that pre-funding of projects could be accepted in specific circumstances, for long-term, large-scale investments, as long as strict safeguards are provided, with assurances that the pre-funded charges will be actually spent on the provision of aviation services, instead of being used for general purposes or other national priorities.[199]

States or air navigation services providers might be tempted to abuse their monopoly position to produce excessively high profits from charges levied above the operational costs.[200] The issue gains particular relevance in the context of the GNSS. Although signals-in-space are, for the time being, provided free of charge by the U.S. and the Russian Federation, in the absence of a competitive environment, due consideration must be given to the desirability of a regulatory mechanism to prevent abuse of monopoly power. A recommendation has been adopted by the LTEP in this regard.[201]

States should exercise caution in their national policy on charges for air navigation services, taking into consideration the effect on users. For instance, it might be necessary for air carriers to adjust their fares in order to cope with or absorb increased costs arising from new or higher charges.[202]

In view of the fact that civil aviation users represent only a minor share of satellite navigation users, it has been recommended that they should not pay for more than their fair share of the costs of GNSS provision.[203] In the words of the LTEP, "cost recovery schemes should ensure the reasonable allocation of costs among civil aviation users themselves, and among civil aviation users and other system users".[204]

Finally, it is extremely important that States ensure that revenues from air navigation services charges are applied solely towards defraying the total costs of these facilities and services.[205] Accordingly, States should permit the imposition of charges only for those services which are "provided for, directly related to, or ultimately beneficial for civil aviation purposes".[206] Revenues exceeding the needs of the service provider or which are unnecessarily diverted to purposes other than civil aviation only add to the costs to be ultimately recovered from the users.[207]

An in-depth review of the ICAO's policy, practical guidance and assistance on financial and organisational aspects of airports and air navigation services, as well as the role and responsibilities of the government has been undertaken by the Conference on the Economics of Airports and Air Navigation Services convened by ICAO at the headquarters, in June 2000.

Financing

Cost-benefit Analysis and Business Case

During the discussions in the Air Navigation Services and Economics Panel (ANSEP), it was made clear that the decision of any State as to whether or when it should enter into financial commitments for the implementation of GNSS in its airspace, like any major investment in the CNS/ATM systems, should be preceded by appropriate financial and economic analyses.[208] The main objective therein would be to establish a cost-effective implementation strategy.

Most significantly, a cost-benefit analysis[209] is deemed essential to identify the investment option that best helps maximise net benefits, and serves to demonstrate the financial viability of a planned investment. It could be accompanied by an economic impact survey to assess the overall contribution of air navigation services to the economy of the State, the understanding of which could help increase the political commitment to the transition process to CNS/ATM systems.[210]

A step further could take the form of a detailed business case[211] to be conducted at national and sub-regional or regional levels, as required. In considering the issue, the Rio Conference further recommended that the concept of homogenous air traffic management and major international traffic flows be taken into account.[212]

The demonstration of sound financial management is therefore critical to securing financing for the systems.[213]

Potential Sources of Funds

Potential sources of funds will vary considerably from State to State, and may include the following:

a) contributions from the national government;
b) contributions from foreign governments, including direct loans and specific aid programmes established to promote economic and social development;
c) loans or grants from development banks;
d) the United Nations Development Programme (UNDP);
e) commercial loans from banks, investment houses and other commercial credit institutions;
f) accumulated excess of revenues over costs (depreciation and retained profits from the operation of air navigation services);

g) bonds;

h) equity financing; and

i) leasing.[214]

Particularly regarding retained profits, the ICAO's Policies on Charges recalls that air navigation services may produce sufficient revenues to exceed all direct and indirect operating costs. Thus, a reasonable return on assets might well be provided to contribute towards necessary capital improvements.[215] In this context, the aviation user charges which may be considered as possible methods for financing GNSS include: i) yearly subscription charges per using operator; ii) yearly subscription charges per using aircraft; iii) year/monthly licence fees; iv) charges per flight; v) charges in respect of different phases of flight; vi) charges based on total passenger-kilometres and tonne-kilometres; vii) regular en-route charges; and viii) a combination of the above.[216]

An alternative much debated and recommended by the LTEP is that GNSS services should be considered as an international service for public use providing the necessary guarantees for accessibility, continuity and quality of the services.[217] States would finance GNSS services as they have financed any other public infrastructure, with the result that the general public would have the right, by law, to demand that efficient services be provided under reasonable charges.[218]

Alternative Mechanisms

ICAO Objectives Implementation Mechanism
The new ICAO policy on technical co-operation, namely the ICAO Objectives Implementation Mechanism, is also strategically linked to the CNS/ATM systems, in the sense that it mobilises additional resources for ICAO follow-up on its Regular Programme activities, which could be applied to Technical Co-operation projects. It is therefore to give priority and support to the implementation of SARPs and air navigation plans, including the CNS/ATM Global Plan.[219]

Until last decade, the funding of projects was almost entirely covered by the UNDP. Nowadays, ICAO acts as a non-profit entity linking both donors and recipient States. Participating States and financing institutions may choose to contribute:

a) for a general fund, which would not be tied to projects for any special area or purpose, nor would have to be used for the purchase of equipment in the donor State or employment of its nationals;

b) for a specific ICAO project;
c) for a specific State project;
d) for a general but identified issue, leaving the manner in which the funds will be spent to ICAO's judgement;[220] or
e) in the form of voluntary contributions in kind, such as scholarships, fellowships, training equipment, and funds for training.[221]

Accordingly, ICAO can assist States in identifying suitable donors for their projects, as well as in the negotiations of convenient funding arrangements.[222] In addition, assistance can be provided concerning the selection of equipment, equipment manufactures, consultants as well as training establishments to meet project goals in the most cost-effective manner.[223] Particularly, the Organisation supports each individual technical co-operation project with its expertise and intimate knowledge of SARPs, and helps ensure the project's sustainability by remaining available for consultation long after its termination.[224]

Therefore, the cost-effectiveness of the services provided through ICAO's Technical Co-operation Bureau, along with the objectivity, impartiality and neutrality of the Organisation are important advantages to be considered by States when seeking for external assistance to implement CNS/ATM related civil aviation projects.[225]

International Financial Facility for Aviation Safety
A proposal for the establishment of an international aeronautical fund was presented by LACAC to the 31st Session of the ICAO Assembly.[226] One amongst the many potential uses suggested was the financing of the implementation of the CNS/ATM systems, under more flexible and less onerous conditions to individual States, where such funding could not be realised through traditional means of cost recovery.[227]

The legal basis for the evolution of the fund derives from ICAO's mandate under Article 44 of the Chicago Convention to ensure the safe and orderly growth of international civil aviation throughout the world, which mandate is further elaborated under Chapter XV.[228]

In subsequent work conducted by ICAO, a study[229] on the viability and usefulness of the establishment of such a fund developed the concept of an International Financial Facility for Aviation Safety (IFFAS) which would have, as an over-riding priority, the financing of safety-related projects for which States could not provide or obtain the necessary financial resources. It has been decided by the ICAO Council that participation by States in any such fund to be developed should be voluntary. Moreover, in order to be considered a potential beneficiary, a State must directly contribute or otherwise participate in the IFFAS.[230]

Three main areas of application have been identified, namely:

a) shortcomings and deficiencies in air navigation services which might impair safety;
b) deficiencies in equipment and training identified through the ICAO safety oversight audit programme; and
c) capital investment requirements for the implementation of the CNS/ATM systems.[231]

Potential sources of funding, which need not be limited to only one mechanism, have been examined and at first focussed on two principal sources, namely: i) an assessment on a participating State based on an amount equal to the State's contribution to the ICAO budget; and ii) a charge on international passengers and/or freight departing from participating States. There seems to be a strong preference for the second option, having a US$1.00 charge per departing passenger been suggested.[232]

In principle, a charge on international passengers used for safety purposes would not be objectionable to the provisions of Article 15 of the Chicago Convention. Nevertheless, concerns were voiced regarding the general principle of equity, in the sense that passengers would be charged for safety-related projects to be provided in the future and for which they might never receive an individual benefit. Yet, safety of international civil aviation could be regarded as a public good, where benefits would be indivisible in terms of individuals. In this context, a passenger safety-charge levied exclusively for safety-related projects would collectively benefit all ICAO member States and the international civil aviation community.[233]

In fact, as previously stated in this Chapter, pre-funding of long-term, large-scale investments for civil aviation projects may be accepted in specific circumstances. Provided that strict safeguards are in place, as regards transparency as well as assurances that all aviation user charges will be allocated to civil aviation purposes, projects to be financed through an IFFAS would qualify under the new ICAO policies concerning airports and air navigation charges.[234]

It has been argued that international financing institutions such as global or regional development banks, which have provided a major share in the financing of airport and air navigation services improvements, are less likely to contribute directly to the creation of an IFFAS. That is understandable considering that such institutions provide assistance only for infrastructure projects which are economically and technically justified, as well as financially viable, but to which individual countries can provide sovereign guarantees. Nevertheless, in a co-operative relationship with an IFFAS financing a safety-related component, they could continue to be relied upon for the remainder of the project. [235]

Contracting States are being particularly encouraged to consider contributing to the capital of an IFFAS by means of voluntary contributions or by crediting to the fund their share - or any amount thereof - of distributable surplus from the ICAO Programme Budget.[236] On the other hand, bequests to the United Nations and a charge on international air cargo are also being investigated, but do not seem to constitute a predictable or sufficient source of fund for the envisaged institution.[237]

In brief, opinions have greatly differed as to the need and appropriateness of establishing an international aeronautical fund. While, for special economic and financial circumstances, the developing world sees countless advantages and benefits to be accrued therefrom,[238] the industrialised nations were at first reluctant and seemed to find insurmountable legal and administrative obstacles to the acceptance of such a fund by their administrations.[239] Their greatest concern was to protect their industry's interests in the maintenance of the common practice of procuring equipment in the donor State. However, having the envisaged fund changed its focus to the financing of safety-related projects, developed nations are slowly beginning to accept the concept of an IFFAS. The European Union has been the first to manifest itself in this regard, and the U.S. is expected to follow suit, at least when IFFAS is already operational.

On the other hand, IATA argues that no passenger or aircraft operator will be willing to accept any more charges,[240] and that air navigation service providers would be guaranteed to recover from aircraft operators all money spent on the provision of the global system, provided States ensure that revenues from airports and air navigation service charges are applied solely towards defraying the costs of these facilities and services.[241] Reality, however, indicates that the reason for IATA's opposition is the fact that an IFFAS will take away its additional income and the influence it currently exerts through the management of user charges for States.

Views have been expressed that the fund could be started at a national or regional level. Accordingly, whereas a framework of concerted principles, guidelines and operating rules for an IFFAS should be defined at a global level, its implementation is recommended to be left at the discretion and initiative of regional groups of States, in order to gather the necessary political support. In this context, it is expected that any involvement by ICAO on the administration of an IFFAS should be upon request of participating States only, and on a cost recovery basis, so as not to make any imposition on the ICAO Programme budget, from which IFFAS shall be completely independent.[242]

The ICAO Council has decided, without objection, to recommend to the Assembly that it endorse the concept of an IFFAS.[243] A draft Assembly Resolution was approved by the Council at its 163rd Session and will be presented, for adoption, to the 33rd Session of the Assembly in September, 2001. In endorsing the concept, the Assembly shall request the Council to pursue the establishment of IFFAS as a matter of priority in the 2002-2004 triennium, and to develop an appropriate implementation mechanism.[244]

In addition thereto, the Council has particularly highlighted the close relationship between the concept of an IFFAS and the Global Aviation Safety Plan, in that both IFFAS and GASP give first priority towards achieving improvements in aviation safety.[245]

As already seen in the previous Chapter, GASP is the umbrella document for all safety-related activities of ICAO, by means of which causative elements of worldwide concern and likely to impair safety are identified, and corresponding safety improvement actions are recommended. Moreover, as an element of the GASP, the ICAO Universal Safety Oversight Audit Programme identifies safety-related shortcomings and deficiencies, and recommends remedial actions.[246]

In order for States to meet their safety oversight responsibilities, it is their responsibility to implement corrective action plans through technical co-operation projects or government programmes. However, many States find difficulties in fulfilling such tasks, mainly due to budgetary constraints. As a result, an IFFAS, which has as its main objective the financing of safety-related projects for which States cannot provide or obtain the necessary financial resources, might constitute an ideal source of funding.

Finally, funding of investments in airports and air navigation services infrastructures, including the CNS/ATM systems, which are expected to remedy many safety-related issues in the long-term future, will most certainly be considered by participating States in an IFFAS.

In fact, capital investments requirements for the implementation of CNS/ATM systems in the developing world are estimated at 3.7 billion dollars. A significant part thereof could be expected to be provided by an IFFAS, although States would still need additional funds to maintain cash flow until cost recovery through user charges begins.[247]

Future Operating Structures

ICAO's policy on future operating structures for GNSS establishes an evolutionary institutional path, best described in the Council Statement of 1994. Accordingly:

> The global navigation satellite system (GNSS) should be implemented as an evolutionary progression from existing global navigation satellite systems, including GPS and GLONASS, towards an integrated GNSS over which Contracting States exercise a sufficient level of control on aspects related to its use for civil aviation. ICAO shall continue to explore, in consultation with Contracting States, airspace users and service providers, the feasibility of achieving a civil, internationally controlled GNSS.[248]

In this regard, the FANS (Phase II) Committee considered a number of institutional options which would provide acceptable GNSS service in accordance with RNP requirements, provided the respective institutional issues were resolved and safety standards were satisfied.[249] These options are:

a) GPS or GLONASS;
b) GPS and GLONASS;
c) GPS/GLONASS plus overlay;
d) GPS/GLONASS plus several civil satellites; and
e) Civil GNSS satellites. [250]

Any option could be selected by a State, subject to its own institutional requirements and the cost-effectiveness of moving on to the next one in the evolutionary path.[251]

Associated implications, which were also considered by the Committee, include the complex issues of operation, ownership and control. It was agreed that irrespective of who owns or operates the space segment, the interests of a State would be served by the institutional options which provide an acceptable level of control to the ATS authority.[252] Accordingly:

> As long as State ATS authorities have control over issues which influence their basic activities such as safety, long and short-term continuity, management, liability, accountability, costs and procurement, every stage in the evolutionary path from GPS and/or GLONASS to a civil GNSS system can be made institutionally acceptable.[253]

The required level of control may vary from State to State, and must be achieved through institutional arrangements, such as:

a) agreements between the GNSS provider and an individual State;
b) agreements with the GNSS provider by a group of States;
c) agreements with an inter-governmental organisation; and
d) joint-support arrangements within the framework of Chapter XV of the Chicago Convention.[254]

A multinational structure would ideally resolve the issue of control were it not for the complex and time-consuming process for setting up a new international organisation for operating the GNSS on behalf of the international civil aviation community. In this respect, a recommendation by the LTEP provides that "to the extent possible, the future systems should make optimum use of existing organisational structures, modified if necessary, and should be operated in accordance with existing institutional arrangements and legal regulations".[255]

The LTEP has further recommended that a centralised operating structure is not needed at this stage. It may, however, be the subject of further study. Meanwhile, national and regional operating structures should be developed. International co-ordination can be achieved through regional organisations operating under the umbrella of ICAO. The Organisation should retain its co-ordinating role with respect to the future GNSS, including the system providing the primary signals-in-space.[256]

An exclusively civil, internationally controlled GNSS remains the ultimate goal in the evolutionary institutional path for the future GNSS.[257] Its feasibility, however, will be dictated by the financial means and the political will of the international community. Thus quite some time may still have to pass before that can be effectively accomplished.[258]

An assumption can therefore be made that the future GNSS will be the result of the evolution of the existing systems, and particularly, the integration of the elements now available with any new ones that might follow. It is not expected to be a single system, but a cluster of different global and regional systems, either civilian-controlled, military-controlled, or a combination of both.[259]

As technology evolves to support the needs of the international civil aviation community, and the navigation satellite systems assume the role of an international asset, a broader acceptance of these services is developing. Still, a globally acceptable system will be one to adequately answer the institutional challenges posed by the GNSS. Particularly, it will have to balance the interests of provider and user States, and to provide a sufficient degree of international control so that the necessary confidence is developed for States to be able to reap the many valuable benefits thereupon.[260]

Notes

1   See J.Huang, "Development of the Long-Term Legal Framework for the Global Navigation Satellite System" (1997) XXII:I *Annals of Air and Space Law* 585 at 593 [hereinafter Huang].

2   See ICAO, *Assembly, 32nd Session, Legal Commission, Recommendations of LTEP*, ICAO Doc.A-32-WP/24 [hereinafter *LTEP Recommendations*], Recommendation 1.

3   See ICAO, *Report of the Panel of Experts on the Establishment of a Legal Framework with regard to GNSS*, ICAO Doc.LTEP/1 (23 December 1996) at para.4:1.10 [unpublished][hereinafter *LTEP/1 Report*]; ICAO, *Report of the First Meeting of the Working Group on GNSS Framework Provisions (Working Group II) of LTEP*, ICAO LTEP/2-WP/3 (15 September 1997) at para.1:6 [unpublished] [hereinafter *WG/ II Report*]; ICAO, *Working Group on GNSS Framework Provisions (Working Group II) of LTEP*, LTEP-WG/II (22-25 April 1997) [hereinafter LTEP-WG/II], "Legal Aspects of GNSS Certification", ICAO Doc.LTEP-WG/II-WP/2 (18 March 1997) at para.6.1.

4   *LTEP 1 Report, ibid.* at para.3:21.

5   M.Milde. "Solutions in Search of a Problem? Legal Aspects of the GNSS" (1997) XXII:II *Annals of Air and Space Law* 195 at 203-204 [hereinafter Milde].

6   See *LTEP 1 Report, supra* note 3 at para.3:24.

7   See Milde, *supra* note 5 at 203.

8   See *LTEP Recommendations, supra* note 2, Recommendation 2; *WG/II Report, supra* note 3 at 1:16. See above, Chapter 5, GNSS SARPs.

9   See ICAO, *Legal Committee, 29th Session*, Report of the Rapporteur on the "Consideration, with regard to global navigation satellite systems (GNSS), of the establishment of a legal framework", by Kenneth Rattray, LC/29-WP/3-1 (3 March 1994) at para.9.

10   See *LTEP 1 Report, supra* note 3 at para.3:22.

11   See *ibid.* at paras.3:25, 3:26; *WG/II Report, supra* note 44, ch. 5 at 1:14.

12   See *LTEP Recommendations, supra* note 2, Recommendation 8.

13   The general understanding is that the term is designed to cover situations where the signals are provided by an entity or organisation other than States.

14   See *LTEP Recommendations, supra* note 2, Recommendation 3.

15   NOTAM (Notice to Airmen) is a notice distributed by means of telecommunication containing information concerning the establishment, condition or change in any aeronautical facility, service, procedure or hazard, the timely knowledge of which is essential to personnel concerned with flight operation. Groenewege, A., *Compendium of International Civil Aviation*, 2nd ed. (Montreal: IADC, 1998) at 565.

16   See *LTEP Recommendations, supra* note 2, Recommendation 3, 6, 7.

17   See *WG/ II Report, supra* note 3 at para.2:7.

18   See *WG/ II Report, ibid.* at 2:9.

19   See ICAO, *Report of the Panel of Legal and Technical Experts on the Establishment of a Legal Framework with regard to GNSS,* ICAO Doc.LTEP/2 (3 November 1997) at para.2:39 [unpublished].

20   See *ibid.*

21   See ICAO, *Report of the First Meeting of the Secretariat Study Group on Legal Aspects of CNS/ATM Systems,* ICAO SSG-CNS/I-Report (9 April 1999)   at para.3.8.5 [hereinafter *Study Group I Report*].

22   See *WG/ II Report, supra* note 3 at para.2:7.

23   See ICAO, *Second Meeting of the Secretariat Study Group on Legal Aspects of CNS/ATM Systems,* ICAO SSG-CNS/2 (20-21 October 1999) [hereinafter SSG-CNS/2], "GNSS Liability: An Assessment", ICAO Doc.SSG-CNS/I-WP/4 (4 October 1999), by F. Schubert, presented by R.D.van Dam at 13.1[hereinafter Schubert and van Dam].

24   *Convention for the Unification of Certain Rules Relating to International Carriage by Air, 12 October 1929, Schedule to the United Kingdom Carriage by Air,* Act 1932; 22 and 23 Geo.5, ch.36 (entered into force 13 February 1933) [hereinafter *Warsaw Convention*].

25   *Convention on Damage Caused by Foreign Aircraft to Third Parties on the Surface,* 7 October 1952, ICAO Doc.7364 (entered into force 4 February 1958) [hereinafter *Rome Convention*].

26   See Schubert and van Dam, *supra* note 23 at para.13.2.

27   See *ibid.*

28   See *Study Group I Report, supra* note 21 at para. 3.10.2.

29   See *Study Group I Report, ibid.* at para.3.10.3; ICAO, *Report of the Panel of Legal and Technical Experts on the Establishment of a Legal Framework with regard to GNSS*], ICAO Doc.LTEP/3 (9 March 1998) at 1:29 [unpublished].

30   See ICAO, *First Meeting of the Secretariat Study Group on Legal Aspects of CNS/ATM Systems,* ICAO SSG-CNS/I-IP/1 (April 1999) at 3.6; M.B.Jennison, "A Legal Framework for CNS/ATM Systems" (ICAO World-wide CNS/ATM Systems Implementation Conference, Rio de Janeiro, 14 May 1998) at 5.

31   See ICAO, *Panel of Experts on the Establishment of a Legal Framework With Regard to GNSS,* LTEP/2 (6-10 October 1997), "Liability Aspects of GNSS", ICAO Doc.LTEP/2-WP/6 (1 October 1997) at para.3, presented by O.Carel, P.O'Neill, F.Schubert, R.D.van Dam, G.White, F.A.Wister.

32   Schubert and van Dam, *supra* note 23 at para.12.

33   See *LTEP/1 Report, supra* note 3 at para.3:33.

34   See *Study Group I Report, supra* note 21 at paras.3.8.6 and 3.8.8.

35   See ICAO, *Report of the Second Meeting of the Secretariat Study Group on the Legal Aspects of CNS/ATM Systems,* ICAO C-WP/11190 (22 November 1999) at para.2.1.3 [unpublished].

36   See *Study Group I Report, supra* note 21 at paras.3.5 and 3.6. In this regard, the Secretariat Study Group on the Legal Aspects of the CNS/ATM has been working on the assumption that "the long-term GNSS, which will be an evolution of the existing systems, will be composed of different global and regional systems. These systems could be civilian-controlled, military-controlled or a mixture of both. The long-term GNSS will include core elements (primary signals-in-space) and augmentation systems". *Ibid., Conclusions of the Study Group on Legal Aspects of CNS/ATM Systems at its First Meeting,* Attachment C at para.1[hereinafter *SSG I Conclusions*].

37 See *LTEP Recommendations, supra* note 2, Recommendation 9.

38 See ICAO, *Report of the Third Meeting of the Secretariat Study Group on the Legal Aspects of CNS/ATM Systems*, ICAO SSG-CNS/3-Report (23 June 2000) [hereinafter *Study Group III Report*] at para.3.1.8 [unpublished]. See also *LTEP/1 Report, supra* note 3 at para.3:32; Huang, *supra* note 1 at 594; Schubert and van Dam, *supra* note 23, at para.11, note 25.

39 See *LTEP/1 Report, ibid.* at para.3:15; *Study Group I Report, supra* note 21 at para.3.2.8.

40 See F.Schubert, "Réflexions sur la Responsabilité dans le Cadre du GNSS" (1997) 45:180 *Revue Navigation* 417 at 417-418 [hereinafter Schubert]. In a similar situation, Annex 11 to the Chicago Convention provides that "by mutual agreement, a State may delegate to another State the responsibility for establishing and providing air traffic services in flight information regions, control areas or control zones extending over the territory of the former. ... It does so without derogation of its national sovereignty. Similarly, the providing State's responsibility is limited to technical and operational considerations and does not extend beyond those pertaining to the safety and expedition of aircraft using the concerned airspace". *Convention on International Civil Aviation*, 7 December 1944, ICAO Doc.7300/6; UN Doc.15 U.N.T.S.295 (entered into force 4 April 1947)[hereinafter *Chicago Convention*], Annex 11, Air Traffic Services.

41 Central American Corporation for Air Navigation Services (COCESNA).

42 See B.D.K.Henaku, "Legal Issues Affecting the Use of Navigation Systems" (1999) 47:187 *Revue Navigation* 312 at 314.

43 See *LTEP/1 Report, supra* note 3 at para.3:15.

44 See Schubert, *supra* note 50 at 418;; Milde, *supra* note 5 at 202; B.D.K Henaku, *The Law on Global Air Navigation by Satellite: A Legal Analysis of the CNS/ATM System* (AST, 1998) at 137 [hereinafter Henaku].

45 See ICAO, *Air Navigation Services and Economics Panel, Report on Financial and Related Organisational and Managerial Aspects of Global Navigation Satellite System (GNSS) Provision and Operation*, ICAO Doc.9660 (May 1996) at para.2.6.1[hereinafter *ANSEP Report*].

46 See *Study Group I Report, supra* note 21 at 3.8.4.

47 LTEP-WG/II, *supra* note 3, "Liability Aspects of GNSS", ICAO Doc.LTEP-WG/II-WP/7 (18 April 1997) at para.2.5.3, presented by F.Schubert [hereinafter LTEP-WG/II-WP/7].

48 See Schubert and van Dam, *supra* note 23 at para.6.2.

49 As the regulatory authority, a State may, for example, "regulate the use of GNSS services by aircraft on its own register when flying outside its airspace, prohibit the use by that aircraft of any GNSS service, permit that aircraft to use GNSS services when no other service is available or permit the use of a GNSS service to the discretion of the aircraft commander". SSG-CNS/2, *supra* note 23, "From Article 28 of the Chicago Convention to the Contractual Chain Solution", ICAO SSG-CNS/2 Flimsy No.1 (21 October 1999), presented by R.D.van Dam [hereinafter van Dam] at 2.

50 See Schubert and van Dam, *supra* note 23 at para.6.2.

51 *LTEP Recommendations, supra* note 2, Recommendation 10.

52 See Schubert and van Dam, *supra* note 23 at para.6.2.

53 van Dam, *supra* note 49.

54 See *Study Group I Report, supra* note 21 at para.3.8.8.

55 van Dam, *supra* note 49 at 3.

56 See *ibid.*

57 See *Study Group I Report, supra* note 21 at para.3.8.8.

58 See *SSG I Conclusions, supra* note 36 at para.2.

59 See *Study Group II Report, supra* note 35, *Conclusions of the Second Meeting of the Study Group on Legal Aspects of CNS/ATM*, Attachment C at para.4 [hereinafter *SSG II Conclusions*].

60 *SSG II Conclusions, ibid.*

61 *Study Group I Report, supra* note 21 at para.3.10.3.

62 In Australia, air traffic services are provided by Airservices Australia, a statutory authority of the Australian Commonwealth Government, created under the Air Services Act 1995. It is regulated by the Civil Aviation Safety Authority, which is also responsible for approving and certifying air navigation procedures and standards, and enforcing safety standards. Both constitute separate corporate bodies which can sue or be sued in their own name. Particularly relevant, Airservices provides ATS on a contractual basis, having negotiated specialised contracts with the major airlines, all of which do address liability. ATS liability is governed by the law of torts, and in particular, the areas of negligence and personal injury. See SSG-CNS/2, *supra* note 23, "An Overview of the Legal Rules in Australia Applicable to Claims Against ATC", ICAO SSG-CNS/2-WP/7 (20 October 1999), presented by S. Clegg.

63 In Canada, air traffic services are provided by NAV CANADA, a non-profit, non-share-capital corporation, created under Part II of the Canada Business Corporations Act. The legal basis for the commercialisation of the air navigation system was previously established by Parliament in the Civil Air Navigation Services Commercialisation Act, the actual sale and transfer having taken effect on 31 October, 1996. Like any private sector entity, NAV CANADA is subject to civil and criminal law. Therefore, in case of damage, the law of the province where the negligent act is alleged to have been committed shall apply. The law of torts will generally apply with the exception of the Province of Quebec, where the general rules on civil liability is found in Article 1457 of the Civil Code of Quebec. See SSG-CNS/2, *supra* note 23, "The Law in Canada", ICAO SSG-CNS/2-WP/3 (4 October 1999), presented by G.Lauzon [hereinafter Lauzon].

64 In France, ATC is operated by the "Direction de la Navigation Aérienne" (DNA), which is under the control of the Ministry of Transportation. The singularity of the French law reveals that an action for damages against the State shall be brought before an administrative court. Moreover, at the option of the plaintiff, it can also be brought before a civil court against an agent or employee of the DNA, who may be personally held liable if it rests proved that he or she committed a personal fault which caused or contributed to the damage. Administrative (case law) and civil (civil code) liability rules will respectively apply. See SSG-CNS/2, *ibid.*, "The Liability System of the French Air Traffic Control", ICAO SSG-CNS/2-WP/2 (20 October 1999), presented by J. Courtial.

65 In Italy, air traffic services are provided by the E.N.A.V., an Economic Public Authority. According to the rules of the Italian Civil Code, an action for damages shall be brought before the civil court. E.N.A.V. will be held liable if the plaintiff proves that a causal nexus exists between the damage and E.N.A.V. activities. Liability is unlimited

and based on fault. See SSG-CNS/2, *ibid.*, "Legal Rules in Italy Applicable to ATC", ICAO SSG-CNS/2-WP/5 (15 October 1999), presented by E.Chiavarelli.

66   In the United Kingdom, air traffic services are provided by the Civil Aviation Authority (CAA) through the National Air Traffic Services (NATS). Section 72 (3) of the Civil Aviation Act 1982 provides that "without prejudice to any right of action in respect of an action or omission which takes place in the course of providing air navigation services ... , no action shall lie in respect to a failure by the CAA to perform [its] duty". Consequently, NATS may be held liable "if, in the course of providing air navigation services, it causes loss or injury as a result of a negligent act or omission. ...Similarly, the CAA, as the aviation safety regulator, is liable for loss or damage resulting from any negligence in carrying out its duties". SSG-CNS/2, *ibid.*, "Legal Rules in the United Kingdom Applicable to ATC", ICAO SSG-CNS/2-WP/1 (20 October 1999), presented by D.J.A.Stoplar [hereinafter Stoplar].

67   In the U.S., the air traffic control system it is operated by the FAA. Suits for damages against the government must be brought before Federal Courts, which will apply the substantive state (tort) law of their jurisdictions. Thus, it is possible that the law on liability differs substantially from one State or Federal Court to another. However, a common Federal code of procedural rules applies to suits in all Federal Courts. In tort claims, proof of negligence alone is not enough to justify recovery. The claimant must establish that there was a breach of a duty of care by the defendant and that the breach of that duty was the proximate cause of the damages. See SSG-CNS/2, *ibid.*, "U.S. Rules for Claims Against Air Traffic Control for Damages or Injury Resulting from Failure of Navigation Aids", ICAO SSG-CNS/2-WP/6 (15 October 1999), presented by M.B.Jennison [hereinafter Jennison].

68   See *SSG II Conclusions, supra* note 59 at 1.

69   *SSG II Conclusions, ibid.*, at 2.

70   Stoplar, *supra* note 66 at 3.

71   Lauzon, *supra* note 63 at 6.

72   Jennison, *supra* note 67 at 3,4.

73   U.S., *The Federal Tort Claims Act*, 28 U.S.C., ss. 1346, 1402, 2401-2415, 2671-2680 (1988)[hereinafter FTCA].

74   FTCA, *ibid.*, .§1346 (b). Moreover, "the U.S. shall be liable ... in the same manner and to the same extent as a private individual under like circumstances". *Ibid.*, .§ 2674".

75   *Indian Towing Co. v. U.S.*, 350 U.S. 61 (1955).

76   See Jennison, *supra* note 67 at 5.

77   See *Delta Air Lines v. U.S.*, 561 F.2d 381 (1st Cir.1977).

78   See, for example, *Hays v. U.S.*, 899 F.2d 438 (5th Cir.1990), where the court found the U.S. 55 per cent negligent and the pilot 45 per cent negligent and awarded damages accordingly. In particular, the FAA inspector in charge of test flights breached the duty to conduct the test with due care for the safety of the aircraft and its passengers and proximately caused injuries. The pilot also had the duty to ensure the safety of flight, which he breached, and his failure was proximate cause of the injuries. See especially, Hamalian, S.K., "Liability of the U.S. Government in Cases of Air Traffic Controller Negligence" (1996) XI *Annals of Air and Space Law* 58 at 65-83, for a case law analysis of liability for the different phases of flight, regarding the pilot-in-command and the ATC. See also for the distribution of liability between the pilot and the ATC,

F. P.Schubert, "Pilots, Controllers, and the Protection of Third Parties on the Surface" (1998) XXIII *Annals of Air and Space Law* 185 at 185 ff.

79   FTCA, *supra* note 73, §2680 (a) [emphasis added].

80   See G.E.Michael, "Legal Issues Including Liability Associated With the Acquisition, Use and Failure of GPS/GNSS" (1999) 54:2 J. Navigation 246 at 247 [hereinafter Michael]. See *Dalehite v. U.S.*, 346 U.S. 15 (1953); *Eastern Airlines, Inc. v. Union Trust Co.*, 221 F.2d (D.C.Cir 1955), where the negligent omission of the control tower operators to issue timely warning to either passenger plane or military plane that the other was on final approach was found by the court not to be a decision responsibly made "at the planning level". Those were merely operational details which are outside the scope of the discretionary function. Consequently, the Government was held liable for damages sustained because of the operators' negligence.

81   *Berkovitz v. U.S.*, 486 U.S. 531 (1988) [emphasis added]. But see *U.S. v. S.A. Empresa de Viação Aérea Rio Grandense (Varig Airlines)*, 467 U.S. 797 (1984), where the Supreme Court extended the discretionary function exception beyond the policy-making level and held that the FAA's acts in executing the "spot check" compliance programme in accordance with agency directives were discretionary and therefore protected. See especially, M.E.F.Plave, "*U.S. v. Varig Airlines*: The Supreme Court Narrows the Scope of Government Liability under the Federal Tort Claims Act". (1985) 51 *Journal of Air Law and Commerce* 198 ff.

82   *Ibid.*, at 536.

83   See K.K. Spradling, "The International Liability Ramifications of the U.S. NAVSTAR Global Positioning System" (1990) 33 *Colloquium on the Law of Outer Space* 93 at 95 [hereinafter Spradling]; P.A.Salin, "An Update on GNSS Before the Next ICAO Experts Meeting on the Legal and Technical Aspects of the Future Satellite Air Navigation Systems" (1997) XXII-I *Annals of Air and Space Law* 505 at 516.

84   Michael, *supra* note 80 at 247.

85   FTCA, *supra* note 73 at §680 (k).

86   See Spradling, *supra* note 83 at 93.

87   Jennison, *supra* note 67 at 6 [emphasis added].

88   See *In re Paris Air Crash of March 31, 1974,* 399 F. Supp.732 (Cal. 1975).

89   *Smith v. U.S.*, 507 U.S. 197 (1993).

90   See *ibid.* at 204.

91   *Ibid.* at 205.

92   *Ibid.* at 207.

93   *Ibid.* at 206.

94   See P. B. Larsen, "Future GNSS Legal Issues" (Third United Nations Conference on the Peaceful Uses of Outer Space, UNISPACE III, 19-30 July 1999) at VI.

95   See FTCA, *supra* note 73 at §2680 (j).

96   U.S., *The Foreign Claims Act*, 10 U.S.C.A. § 2734 (1996).

97   U.S., *The Military Claims Act*, 10 U.S.C.A. § 2733 (1996).

98   van Dam, *supra* note 49 at 1.

99   See van Dam, *ibid.* at 3. "The principle of the liability of the providing State has been recognised by 28 European States ("Umbrella Agreement on the Delegation of ATS")". *Ibid.* For example, the Special Agreement Relating to the Operation of the Maastricht Control Centre by Eurocontrol, which relates to the provision and operation of en route air traffic facilities and services at the Maastricht Control Centre on behalf of Germany,

Belgium, Luxembourg and the Netherlands, provides that the Organisation is liable for damages arising out of the performance of its task under the agreement. However, it has a right of recourse against any State which is found liable. Likewise, should a National Contracting Party be ordered to make good damage for which the Organisation was liable, the latter would be required to indemnify the State concerned. See LTEP-WG/II, *supra* note 32, "Analysis of Liability Provisions in Existing International Conventions, Treaties and Other Relevant Instruments and Their Applicability to GNSS", LTEP-WG/II-WP/9 (18 April 1997) at para.10, presented by O.Carel, M,Denney, E.Hoffstee, P.O'Neill, T.Nordeng. W. t'Hoen, A.Watt, G.White [hereinafter LTEP-WG/II-WP/9].

100 For more information on the concept of channelling of liability and the contractual framework, see above, Chapter 6 at 105.

101 See LTEP-WG/II, "Legal Aspects of GNSS Certification and Liability", LTEP-WG/II-WP/8 (18 April 1997) at para.3.3.2, presented by O.Carel, M,Denney, E.Hoffstee, P.O'Neill, T.Nordeng., W. t'Hoen, A.Watt, G.White [hereinafter LTEP-WG/II-WP/8].

102 See Canada, *State Immunity Act*, S.C. 1980-81-82-83, c. 95, s. 5, 6. See also, Lauzon, *supra* note 63 at 3, 7.

103 See *SSG II Conclusions, supra* note 59 at paras.2,3; *Study Group III Report, supra* note 38 at para.3.2.6.

104 See Schubert and van Dam, *supra* note 23 at 16.

105 See *ibid.*

106 Schubert and van Dam, *ibid.* at 15.

107 For a comprehensive review of the developments in the field of products liability law, see R.I.R.Abeyratne, "The Evolution from FANS to CNS/ATM and Products Liability of Technology Providers in the U.S." (1994) 43:2 *Zeitschrift sur Luftund Weltraumrecht* 156 at 156-186; G.R.Baccelli, "La Responsabilità del Construttore Aerospaziale Secondo la Giurisprudenza Comparatistica e la Direttiva CEE in Materia di Responsabilità per Prodotto Difettoso (1990) XIV:2 *Diritto e Pratica dell'Aviazione Civile* 359 at 359-366. See especially, EU, *Directive 85/374/EEC of 25 July 1985 on the Approximation of the Laws, Regulations and Administrative Provisions of the Member States Concerning Liability for Defective Products*, [1985] O.J.L. 210/29. See also P.D.Bostwick, "Liability of Aerospace Manufactures: MacPherson v. Buick Sputters into the Space Age" (1994) 22 *Journal of Space Law* 75 at 75-96.

108 Schubert and van Dam, *supra* note 23 at 19.

109 See LTEP-WG/II-WP/7, *supra* note 47 at para.2.7.2.

110 *Warsaw Convention, supra* note 24, Article 17.

111 See *Warsaw Convention, ibid.*, Article 25.

112 *Convention for the Unification of Certain Rules for International Carriage by Air*, 28 May 1999, DCW Doc.No.57 (28 May 1999) (not yet in force)[hereinafter *Montreal Convention*]. For a detailed analysis of the provisions of the Montreal Convention, see A.A.L., Andrade, "Convenção de Montreal: Derradeira Esperança para o Transporte Aéreo Internacional" (1999) 78 *Revista Brasileira de Direito Aeroespacial* 2 at 2-18.

113 *Montreal Convention, supra* note 112, Article 21.

114 See *Warsaw Convention, ibid.*, Article 28.

115 See *Montreal Convention, supra* note 112, Article 33.

116 See *ibid.*, Article 28.

117 See *ibid.*, Article 50.

118 See *ibid.*, Article 53.

119 See *Rome Convention, supra* note 25, Article 1.

120 See *ibid.*, Article 6.

121 See *ibid.*, Article 12.

122 See Schubert and van Dam, *supra* note 23 at 14.

123 See *ibid.* at 15.

124 *Convention on the International Liability for Damage Caused by a Space Object,* 29 March 1972, 961 U.N.T.S. 187 [hereinafter *Liability Convention*], Article II. For a comprehensive review on the Liability Convention, see W.F.Foster, "The Convention on International Liability for Damage Caused by Space Objects" (1972) *Canadian Yearbook of International Law* 137. See also, B.Cheng, "International Responsibility and Liability for Launch Activities" (1995) XX:6 *Annals of Air and Space Law* 297 at 297-310. For more information on the law on liability for damage caused by a space object under the Outer Space Treaty and the Liability Convention, see D.Maniatis, "The Law Governing Liability for Damage Caused by Space Objects" (1997) XXII-I *Annals of Air and Space Law* 369 at 369-401.

125 See S.Gorove, *Developments in Space Law, Issues and Policies* (Dordrecht: Martinus Nijhoff /Kluwer, 1991) at 149; Milde, *supra* note 5 at 212. But see B.A.Hurwitz, *State Liability for Outer Space Activities in Accordance with the Convention on the International Liability for Damage Caused by a Space Object* (Dordrecht, Martinus Nijhoff, 1992) at 31; Henaku, *supra* note 44 at 221 to 233.

126 See *Liability Convention, supra* note 124, Articles IX, XI. Nothing prevents a State from pursuing a claim in the courts or administrative tribunals of a launching State. In this case, however, it shall not be entitled to present a claim for compensation in respect of the same damages under the Convention.

127 See *WG/ II Report, supra* note 3 at para.2:3. See also, Henaku, *supra* note 44 at 237-238.

128 See *LTEP Recommendations, supra* note 2, Recommendation 11(d).

129 See LTEP-WP/II-WP/8, *supra* note 101 at para.5.5.1.

130 See G.R.Baccelli, "L'Unification Internationale du Droit Privé Aérien: Perspectives en Matière de Responsabilité des Transporteurs, des Exploitants des Aéroports et des Services de Contrôle de la Circulation Aérienne" (1983) VIII *Annals of Air and Space Law* 3 at 19-20.

131 See *ibid.* at para.5.5.2.

132 See LTEP-WG/II-WP/9, *supra* note 99 at para.9.

133 *International Convention on the Establishment of an International Compensation Fund for Oil Pollution Damage,* 18 December 1971(entered into force 16 October 1978), as amended 19 November 1976 and 25 May 1984 (not yet into force) [hereinafter *Fund for Oil Pollution Damage Convention*].

134 *International Convention on Civil Liability for Oil Pollution Damage,* 29 November 1969 (entered into force 19 June 1975) as amended 19 November 1976 (entered into force 8 April 1981) and 25 May 1984 (not yet into force) [hereinafter *1969 Liability Convention*].

135 See *Fund for Oil Pollution Damage Convention, supra* note 133, Preamble.

136 *Ibid.*, art. 4, para.1.

137 See *ibid.*, art. 2 (c). Each Contracting State shall recognise the Director of the Fund as the legal representative of the Fund.

138 See *ibid.*, art 4, para.3.

139 See *ibid.*, art. 5.

140 See *ibid.*, arts. 10-12.

141 See *WG/II Report, supra* note 3 at para.2:3.

142 *Operating Agreement on the International Maritime Satellite Organisation (INMARSAT)*, 3 September 1976, 31:1 U.S.T.135 (entered into force 16 July 1979), Article XII [emphasis added] [hereinafter *Operating Agreement*].

143 *Constitution of the International Telecommunications Union*, Geneva, 22 December 1992 (entered into force 1 July 1994), art. 36, No. 183 [hereinafter *ITU Constitution*].

144 See LTEP-WG/II-WP/9, *supra* note 99 at para.2.8; ICAO, *Panel of Experts on the Establishment of a Legal Framework With Regard to GNSS*, LTEP/1 (25-30 November 1996), "Inmarsat Satellite Navigation Services Institutional and Contractual Aspects", ICAO Doc.LTEP/1-WP/11 (29 October 1996) at para.2.11 [hereinafter LTEP/1-WP/11]; Henaku, *supra* note 44 at 220.

145 See *Operating Agreement, supra* note 142, ARTICLE XI. See also LTEP/1-WP/11, *ibid.* Inmarsat maintains insurance cover for claims by third parties, against Inmarsat, its Parties, Signatories and Navigation Land Earth Stations operators. The Organisation shall reimburse the Signatory to the extent that it has paid the claim. The Signatories shall, to the extent that the reimbursement is not satisfied by indemnification, insurance or other financial arrangements, pay to the Organisation the unsatisfied amount of the claimed reimbursement in proportion to their respective investment shares as of the date when the liability arose.

146 See LTEP-WG/II, *supra* note 3, "Liability Aspects of GNSS", ICAO Doc.LTEP-WG/II-WP/3 (18 March 1997) at para.5.4.2.

147 See *LTEP/1 Report, supra* note 3 at para.5.4.4; LTEP/1-WP/11, *supra* note 144 at paras.5.4.2, 5.4.3; *Study Group III Report, supra* note 38 at 3.3.

148 *LTEP Recommendations, supra* note 2, Recommendation 11 b).

149 See above, Chapter 6 at 105.

150 14 in favour, 7 against and 1 abstention.

151 *LTEP Recommendations, supra* note 2, Recommendation 11 bis.

152 See *LTEP/1 Report, supra* note 3 at para.3:34.

153 See *LTEP Recommendations, supra* note 2, Recommendation 9 (i); LTEP-WG/II-WP/7, *supra* note 47 at para.5.3.1.

154 See LTEP-WG/II-WP/7, *ibid.* at paras.5.3.5-5.3.8.

155 See *LTEP Recommendations, supra* note 2, Recommendation 9 (a), (d).

156 See ICAO, *World-wide CNS/ATM Systems Implementation Conference* (Rio de Janeiro, 11-15 May 1998) [hereinafter WW/IMP], "Specific Organisational Aspects Pertaining to the ICAO CNS/ATM Systems", ICAO WW/IMP-WP/15 (4 February 1998) at para.1.1.

157 See ICAO, *World-wide CNS/ATM Systems Implementation Conference Report*, ICAO Doc.9719 (May 1998) at paras.2.1.1, 2.2.1 [hereinafter *WW/IMP Report*].

158 See ICAO, *Global Air Navigation Plan for CNS/ATM Systems,* version 1 (Montreal: ICAO, 1998), vol. 1 at para.12.2.1 [hereinafter *Global Plan*].

159 See generally, WW/IMP, *supra* note 156, "International Co-operative Ventures", ICAO WW/IMP-WP/17 (25 February 1998) [hereinafter WW/IMP-WP/17].

160 See G.Finnsson, "Airports and Route Facilities: International Cost Recovery Policies and Their Applicability in the Framework of New Forms of Infrastructure Provision" (1994) XIX:II *Annals of Air and Space Law* 283 at 291[hereinafter Finnsson].

161 See WW/IMP, *supra* note 156, "Organisational Forms of Air Navigation Services at the National Level", ICAO WW/IMP-WP/16 (6 February 1998) at para.3.2 [hereinafter WW/IMP-WP/16].

162 See T.R.Kesharwani, "Privatisation in the Provision of Airport and Air Navigation Services" (ICAO Airport Privatisation Seminar, Forum for the NAM/CAR/SAM Regions, Guatemala City, 13 December 1999).

163 See Finnsson, *supra* note 160 at 292.

164 See WW/IMP-WP/16, *supra* note 161 at para.4.1.

165 See *ibid.* at 4.1, 4.2.

166 See *WW/IMP Report, supra* note 157, Recommendation 2/4.

167 See Finnsson, *supra* note 160 at 293.

168 See ICAO, *Conference on the Economics of Airports and Air Navigation Services* [hereinafter *ANSConf 2000*], "Study on Privatisation in the Provision of Airports and Air Navigation Services", ICAO ANSConf-WP/6, Appendix A.

169 For detailed information on the commercialisation of the Canadian air navigation system, see D.T.E.Mein, "La Commercialisation du Système de Navigation Aérienne du Canada" (1998) 46:184 *Revue Navigation* 477 at 474-486.

170 See WW/IMP-WP/16, *supra* note 161 at para.5.1.

171 See, ICAO, *Report of the Conference on the Economics of Airports and Air Navigation Services* [hereinafter *ANSConf Report*], ICAO ANSConf-WP/115 (19 – 28 June 2000), Recommendation 2/2. The Conference further recommended that when establishing an autonomous authority, the State should stipulate as a condition for its approval that it observes the obligations of the State specified in the Chicago Convention, as well as other ICAO policies and practices. See *ibid.*, Recommendation 4/1. See also K.O.Rattray, "Economic Regulation – What is the Role of Government in a Changing World?" (Conference on the Economics of Airports and Air Navigation Services, Montreal, 19 June 2000), ICAO ANSConf-SP/7.2.

172 See *WW/IMP Report, supra* note 157, Recommendation 2/7.

173 See *Global Plan, supra* note 158, vol. 1 at para.12.4.2.2; WW/IMP-WP/17, *supra* note 159 at para.2.1ff.

174 For practical examples of charges collection services provided by Eurocontrol, the United Kingdom and IATA on behalf of, respectively, Eurocontrol's member States, DEN/ICE Joint Financing Agreements, and a number of developing States, see ICAO, *ANSConf 2000, supra* note 168, "Joint Ventures in the Billing and Collection of Air Navigation Services", ICAO ANSConf-WP/22 (7 April 2000) at 4.

175 See *Global Plan, supra* note 158, vol. 1 at para.12.4.1; WW/IMP-WP/17, *supra* note 159 at para.4.1ff; Finnsson, *supra* note 160 at 298-302.

176 See G.F.FitzGerald, "ICAO and the Joint Financing of Certain Air Navigation Services" – Part I (1986) XI *Annals of Air and Space Law* 17 at 19.

177 See *Chicago Convention, supra* note 40, art. 69.

178 See *ibid.*, art. 70.

179 See *ibid.*, art. 71.

180 *Ibid.*, art 73.

181 See *ibid.*

182 See *ibid.*, art. 74.

183 See *Global Plan, supra* note 158, vol. 1 at para.12.2.4.2.

184 See *LTEP Recommendations, supra* note 2, Recommendation 13 (2).

185 *Agreement on the Joint Financing of Certain Air Navigation Services in Greenland and the Faroe Islands*, 1956, ICAO Doc.7726-JS/563, *Agreement on the Joint Financing of Certain Air Navigation Services in Iceland*, 1956, ICAO Doc.7727-JS/564 [hereinafter DEN/ICE Agreements].

186 See *Global Plan, supra* note 158, vol. 1 at para.12.2.4.3.

187 See WW/IMP-WP/17, *supra* note 159 at para.5.2.

188 See G.F.FitzGerald, "ICAO and the Joint Financing of Certain Air Navigation Services" - Part II (1987) XII *Annals of Air and Space Law* 33 at 40-41[hereinafter FitzGerald].

189 See FitzGerald, *ibid.* at 46, 50.

190 See WW/IMP-WP/17, *supra* note 159 at para.5.6.

191 See *ibid.* at paras.5.3, 5.4, 5.5.

192 Air navigation services charges are specifically related to such services, and include route, approach and aerodrome control charges. Airport charges comprise, *inter alia*, landing, terminal, hangar, passenger services, parking, security and noise charges. See R.Janis, "Charges vs. Taxes – Which is which and why?" (Conference on the Economics of Airports and Air Navigation Services, Montreal, 19 June 2000), ICAO ANSConf-SP/3.1.

193 See WW/IMP, *supra* note 156, Recommendation 3/10; *Global Plan, supra* note 158, vol. 1 at para.14.1.1.

194 ICAO, *ICAO's Policies on Charges for Airports and Air Navigation Services*, ICAO Doc.9082/6 (2001) [hereinafter *ICAO's Policies on Charges*].

195 See *ICAO's Policies on Charges, ibid.* at para.36.

196 See *ibid.* at para.47.

197 *Ibid.* at para.38 (i).

198 See *ibid.* at para.38 (ii).See *ANSConf Report, supra* note 171, Recommendation 5.2/8.

199 See *ibid.* at para.42. For more information on the subject, see ICAO, *ANSConf 2000, supra* note 168, "Pre-funding of Projects Through Charges", ICAO ANSConf-WP/15 (16 February 2000); ICAO, *ANSConf 2000, supra* note 711, "Forward Financing", ICAO ANSConf-WP/31(24 December 1999) presented by IATA. For the relation of the pre-funding concept with the International Financial Facility for Aviation Safety (IFFAS), see below at 151.

200 See Finnsson, *supra* note 160 at 289.

201 See *LTEP Recommendations, supra* note 2, Recommendation 13 (1).

202 See WW/IMP, *supra* note 156, "International Cost Recovery Policy", ICAO WW/IMP-WP/23 (2 March 1998).

203 See *Global Plan, supra* note 158, vol. 1 at para.14.2.2.3.

204 *LTEP Recommendations, supra* note 2, Recommendation 13 (3). See also *ANSConf 2000, supra* note 171, "The Allocation of GNSS Costs", ICAO ANSConf 2000-WP/65 (12 May 2000) presented by Eurocontrol, where a method for allocating GNSS costs, based on users' requirements, is proposed, namely between i) civil aviation and other categories of users; ii) States; and iii) phases of flight.

205 See *WW/IMP Report, supra* note 157, Recommendation 3/11.

206 See *ANSConf Report, supra* note 171, Recommendation 5.2/13.

207 See *ANSConf 2000, supra* note 168, "Revenue Diversion", ICAO ANSConf-WP/77 (5 June 2000), presented by IATA.

208 See *ANSEP Report, supra* note 45 at para.3.2.1.

209 For further guidance on cost-benefit analysis, *see* ICAO, *Economics of Satellite-Based Air Navigation Services – Guidelines for Cost/Benefit Analysis of Communications, Navigation, Surveillance/Air Traffic Management (CNS/ATM) Systems*, ICAO Circ.257-AT/106. For a practical example, see Spain, *CNS/ATM Cost-Benefit Analysis for Spain: Final Report* (Aeropuertos Españoles y Navegación Aérea, 1996) vol. 1,2. See S.Draghi, "Estimation du Coût de la Mise en Oeuvre du GNSS en Europe" (1996) 44:173 *Revue Navigation* 25 at 25-40, for an analysis of a cost estimation for the implementation of GNSS in Europe conducted by Eurocontrol. See also D.Diez and M.Nárdiz, "Un Estudio de Rentabilidad Sobre el CNS/ATM Realizado Por España Da Resultados Positivos a Nivel Nacional" (1998) 8:40 *Boletín Informativo Aital* 12 at 12-13, for the results of a cost/benefit analysis for the implementation of CNS/ATM systems in Spain.

210 See *Global Plan, supra* note 158, vol. 1 at para.14.3.3.

211 A business case could be defined as "a study that includes the analyses of both costs and benefits of CNS/ATM systems implementation options and the requirements for a financing scheme, including revenues, expenses, and pay back periods". ICAO Secretariat, *Transition, ICAO CNS/ATM Newsletter* 98/05, "Business Cases Essential to CNS/ATM Systems Planning" (Autumn 1998) at 2. See D.L.Allen, A.Haraldsdottir, R.W.Lawler, K.Pirotte, R.Schwab, "The Economic Evaluation of CNS/ATM Transition" (1999) 47:185 *Revue Navigation* 25 at 25-50, for an example of a methodology supporting business case development.

212 See *WW/IMP Report, supra* note 157, Recommendation 3/10.

213 See *ibid.*, Recommendation 3/6.

214 See especially, WW/IMP, *supra* note 156, "Sources of Funds and Financing Mechanisms", ICAO Doc.WW/IMP-WP/25 (2 March 1998).

215 See *ICAO's Policies on Charges, supra* note 194 at para.38 (iv).

216 *LTEP Recommendations, supra* note 2, Recommendation 14.

217 See *LTEP Recommendations, ibid.*, Recommendation 12.

218 See J. Huang, "ICAO Panel of Experts Examining the Many Legal Issues Pertaining to GNSS" (1997) 52:8 *ICAO Journal* 19 at 22 [hereinafter Huang].

219 See *ICAO, Council, 154ᵗʰ Session,* "Study on a Proposal for an International Aeronautical Monetary Fund", Appendix, "The Funding Mechanism of the ICAO Technical Co-operation Programme and Some Examples of Multilateral Mechanisms", ICAO C-WP/10840 (30 April 1998) at para.1.2.

220 See *ibid.* at para.1.3.

221 See WW/IMP, *supra* note 156, "ICAO Objectives Implementation Mechanism and Technical Co-operation", ICAO Doc.WW/IMP-WP/28 (11 May 1998) at para.1.4 [hereinafter WW/IMP-WP/28].

222 See *ibid.* at Appendix A, "Specific Features of the ICAO Objectives Implementation Mechanism of Interèst to States".

223 See *ibid.* at Appendix B, "Specific Features of the ICAO Objectives Implementation Mechanism of Interest to Donors".

224 See R.C.Costa Pereira, "Funding and Implementing Regional and Sub-regional Solutions in Africa" (African Aviation Conference and Exhibition 1999, Washington, 28 June 1999) at 12 [unpublished].

225 See WW/IMP-WP/28, *supra* note 221, Appendix A.

226 See ICAO, *Assembly, 31ˢᵗ Session, Executive Committee*, "Strategic Action Plan", ICAO A31-WP/73.

227 See R.I.R.Abeyratne, "The Latin American Initiative Towards Funding the CNS/ATM Systems" (1998) 77:143 T.A.Q. 151 at 151. See also ICAO Secretariat, *Transition, ICAO CNS/ATM Newsletter* 98/05, "ICAO Examines Establishment of An International Aeronautical Fund" (Autumn, 1998) at 3; *WW/IMP Report, supra* note 157 at 3.3.2.

228 For a description of the ICAO Council's mandate under Chapter XV of the Chicago Convention and its application in the form of joint-financing of air navigation, see above at 144ff.

229 See ICAO, *Council, 158ᵗʰ Session*, "Study on an International Aeronautical Fund", ICAO C-WP/11235 (24 September 1999) [hereinafter *IFFAS Study*]. For the results of a survey conducted by ICAO through a questionnaire to States regarding the study, see ICAO, *Council, 161ˢᵗ Session*, "International Financial Facility for Aviation Safety", ICAO C-WP/11467 (20 October 2000) [hereinafter *IFFAS C-WP/11467*].

230 See *IFFAS C-WP/11467, ibid.* at 3.4. See also ICAO, *Council, 161ˢᵗ Session, 12ᵗʰ Meeting,* ICAO C-DEC 161/12 (30 November 2000) at paras. 2, 4 (b).

231 See ICAO, *Council, 162ⁿᵈ Session*, "Possible Relationship between an International Financial Facility for Aviation Safety (IFFAS) and the Global Aviation Safety Plan (GASP), ICAO C-WP/11545 (6 February 2001).

232 A passenger charge applied to transit passengers, however, would be inconsistent with Article 15 of the Chicago Convention, which provides that no charges shall be imposed solely for the right of transit over the territory of a contracting State of any aircraft or persons thereon. See *IFFAS C-WP/11467, supra* note 299 at 3.5.

233 See *IFFAS Study, supra* note 229 at para.3.2.7.

234 See above, Cost Recovery at 146.

235 See *ibid.* at para.3.4.

236 See ICAO, *Council, 163ʳᵈ Session*, "Draft Assembly Working Paper on the Establishment of a Mechanism for an International Financial Facility for Aviation Safety (IFFAS), Discussion Paper n. 1 related to C-WP/11651 (20 June 2001) at para. 5. See also ICAO, *Council, 163ʳᵈ Session, 19ᵗʰMeeting,* ICAO C-DEC 163/19 (29 June 2001) at para. 2.

237 See *IFFAS Study, supra* note 229 at para.3.5.

238 See M. Folchi, Address (Panel on the Establishment of an International Aeronautical Monetary Fund, Salvador, Brazil. 13 June 1994); A.M.Donato, Address, *ibid.,* J.Razafy, Address, *ibid;* S.A.Al-Ghamdi, "Alternative Approach to Implementation of CNS/ATM Systems Would Impose User Charge" (1993) 48:3 *ICAO Journal* 19 at 19, 20.

239 See ICAO, *Council, 154ᵗʰ Session*, "Study on a Proposal for an International Aeronautical Monetary Fund", ICAO C-WP/10880 (15 May 1998) at para.2.4.

240 See T.Kelly, Address, (Panel on the Establishment of an International Aeronautical Monetary Fund, Salvador, Brazil. 13 June 1994).

241 See ICAO Secretariat, *Transition, ICAO CNS/ATM Newsletter* 98/05, "ICAO Examines Establishment of and International Aeronautical Fund" (Autumn 1998) 3 at 3.

242 See ICAO, *Council, 163ʳᵈ Session*, "Draft Assembly Working Paper - Mobilization of Funds for Civil Aviation", Attachment to C-WP/11591 at 4.1.2 (8 May 2001).

243 See ICAO, *Council, 163ʳᵈ Session*, ICAO C-DEC 163/9 (14 June 2001).

244 See *supra* note 236, ICAO C-DEC 163/19.

245 See ICAO, *Council, 163rd Session,* "Draft Assembly Working Paper - Mobilization of Funds for Civil Aviation - ICAO C-WP/11591 (8 May 2001) at para. 4.3.

246 See *ibid.* See Chapter 6, Global Aviation Safety Plan at 107.

247 See *IFFAS Study, supra* note 229 at para.2.7.

248 ICAO, *Statement of ICAO Policy on CNS/ATM Systems Implementation and Operation,* ICAO Doc. LC/29 - WP/3-2 (28 March 1994) at para.6.

249 See ICAO, *Report of the Fourth Meeting of the Committee for the Monitoring and Co-ordination of Development and Transition Planning for the Future Air Navigation System (FANS PHASE II),* ICAO Doc.9623 - FANS (II)/4 (15 September – 1 October 1993) at para.6.2.5.3 [hereinafter *FANS (II)/4 Report*].

250 See *FANS (II)/4 Report, ibid.,* Appendix A to the Report on Agenda Item 6.

251 See LTEP-WG/II, *supra* note 3, "Future Operating Structures", ICAO Doc.LTEP-WG/II-WP/5 (18 March 1997) at para.2.1.

252 See *FANS (II)/4 Report, supra* note 245 at para.6.3.3.5.

253 *FANS (II)/4 Report, ibid.* at para.6.3.3.7.

254 See *ibid.* at para.6.3.3.6. See especially, Huang *supra* note 1 at 597.

255 *LTEP Recommendations, supra* note 2, Recommendation 15 (3).

256 See Huang, *supra* note 218 at 22; *LTEP recommendations, ibid.,* Recommendation 16.

257 The LTEP has recommended that the future GNSS primary signals-in-space should be civilian-controlled, with user States exercising an appropriate level of control over the administration and regulation of those aspects that relate to civil aviation.

258 See *FANS (II)/4 Report, supra* note 245 at para.6.5.5.

259 See *Study Group I Report, supra* note 21 at para.3.6.

260 See *FANS (II)/4 Report, supra* note 245 at para.6.5.2.

Conclusions

In light of the current interpretation of the Chicago Convention, and as a direct consequence of the principle of sovereignty, a contracting State is indeed ultimately responsible for the provision of air navigation services in its sovereign territory.[1]

In this sense, it is responsible to the international civil aviation community for guaranteeing that the services and facilities provided, whether or not delegated, in total or in part, to another private or multinational entity, to an autonomous authority or to a foreign State or entity, comply with the established international standards applicable in its territory.

Delegation of service provision to a third party is a possibility arising out of the absence of any reference in the Chicago Convention to a specific mechanism by which a State shall fulfil its obligations under Article 28. However, the act of delegation embraces only the actual operational performance of these services. The reason is that a State cannot - ever – release itself from its responsibility, as the sovereign regulatory authority, for the promulgation and enforcement of safety regulations in its territory.

Accordingly, not only does the State retain its responsibility for setting and maintaining the standards, but also, and particularly, for the quality of the services provided.[2] Therefore, when authorising the use in its airspace of services provided by any third party, it must primarily satisfy itself that they are in accordance with ICAO SARPs. The State may be said to have yet another level of responsibility, supervisory in nature, in the sense that it must continuously monitor service compliance with the applicable standards. Through the enactment of appropriate national legislation and the institution of enforcement mechanisms, it must provide the definition by law of the penalty to be applied in the event of their infringement or violation. For example, a situation may arise where it may be held liable by the failure to regulate, or by a faulty exercise of its regulatory power, or else for not having exercised its oversight function with reasonable care, in accordance with its national law.

Still, delegation regularly comes accompanied by the appropriate stipulation of contractual terms and conditions, whereby the delegating State and the service provider both safeguard their interests through the proper allocation of rights and duties, and respective liabilities. Consequently, although there is always a possibility that a State may be directly and solely held liable in case of damage by a binding court decision, a recourse action against the entity providing the services is facilitated by the existence of such a contractual relation.

In practical terms, when entering the airspace of an Article 28 State, it does not matter to the user, namely aircraft engaged in international air navigation from other contracting States, which entity happens to be *in control* of any particular element of the air navigation infrastructure, but only that the State *guarantees* that services are provided within the required level of safety, and in accordance with ICAO SARPs. Any private arrangements which the State might have entered into with the entity providing the services, partially or totally releasing its responsibility for the performance of the services, shall not affect the State's ultimate responsibility under Article 28 for providing the users with the necessary guarantees.

The introduction of satellite-based air navigation, in particular the Global Navigation Satellite System, by no manner of means modifies the obligations of States under Article 28, as described above. The reason is, and again, that the mechanisms by which a State may fulfil such obligations have not been prescribed by the Convention.[3] Accordingly, no State is obliged to make use of satellite technology as an aid to air navigation in its sovereign airspace, and cannot be held responsible under the Convention unless it has *expressly authorised* its use. Therefore, as far as the GNSS is concerned, although it is perfectly proper that a State uses the services of a foreign provider of signals-in-space for providing air navigation services in its territory, the State has to specifically authorise the use of the signal-in-space, through a regulatory act, as well as to continuously monitor its compliance with applicable standards.

Yet, it is undeniable that "the GNSS represents a dramatic step away from past practice in the application of the principle of sovereignty".[4] Whereas States have traditionally retained full control over all elements of their air navigation infrastructure, "the GNSS facilities, at least as far as the space segment is concerned, will be controlled and operated by one or more foreign countries",[5] and therefore no longer under the control of the State undertaking responsibility under Article 28. The controversial issue of control gains particular importance in the perspective of having the GNSS approved as the sole-means of navigation in the State's sovereign territory.

In this regard, legal arrangements whereby a link is established between the provider of signals-in-space and the user State, with the appropriate delegation of duties, are unquestionably necessary to deal with the disparity between responsibility and loss of control, and allow for the proper allocation of liabilities.[6]

Thus having been said, the real situation can be depicted as follows:

Signals-in-space have been offered free of direct user charges to the international community by the governments of the U.S. and the Russian Federation.[7] "The signal is up there" and States may choose to incorporate it in their respective air navigation infrastructure, approving aircraft operations based on its use. In doing so, *they do it of their own free will*, no "formal" legal guarantees having ever been offered by the provider States as regards the availability, continuity, accuracy, reliability and integrity of the GPS and GLONASS systems.

A point of attention must be drawn here as to the fact that these are exactly the same guarantees which the State is legally obliged to provide to international users as regards the RNP requirements for the services provided in its airspace, and may be held accountable for in case of a GNSS related accident. But how can any State possibly guarantee the quality and safety of services it does not control, and has no means of enforcing safety regulations thereupon, or ensuring that applicable international standards will be complied with?

In this regard, SARPs alone cannot be considered sufficient to build up the necessary confidence in the integrity of the system. Moreover, SARPs provide only technical assurances for certified systems and cannot address the necessary liability issues.[8]

On the other hand, the controversy pertaining to the legal significance of the exchange of letters reveals that these instruments do not constitute formal international agreements, nor was there ever any intention on the part of the U.S. or the Russian Federation to make them legally binding.[9] Otherwise, proper internal procedures for entering into executive agreements, which "are to all intents and purposes binding treaties under international law",[10] would have been followed, there being a clear distinction between such agreements and mere unilateral policy statements, not enforceable in law.

Although it has been continuously alleged that "these agreements may be considered morally and politically binding by the parties, and the President may be making a type of national commitment when he enters one",[11] the blunt fact is that presidents change, and policy directives are not eternal.

This is not to say that they are both military systems, and therefore of paramount importance to national security. For example, GPS has been integrated into virtually every facet of U.S. and allied military operations, which are increasingly reliant on its signals for a variety of purposes, from navigation to modern precision-guided weapons and munitions.[12]

How can the international community risk sole reliance on the good faith of the signal-in-space provider State, when knocking at its doorstep is the perpetual danger of having the accuracy of the signal selectively degraded for national security reasons, or abusive user charges imposed for alleged financial constraints, or even a complete shut-down of the entire systems for whatever reasons? In these conditions, which country is prepared to approve the use of GNSS as the sole means of navigation in its territory?[13]

Genuine concerns of States border on the imminence of important financial and budgetary decisions regarding the implementation of the CNS/ATM systems, in which an eventual and progressive withdrawal of current air navigation systems is envisaged. However, in the absence of any other legal or institutional guarantees, redundancy in air navigation facilities, namely an automatic switch to a back-up system on stand-by in case of malfunctioning, might rest as the only practical remedy. Now, a question has to be raised as to the cost-effectiveness, if any, of implementing and maintaining two parallel air navigation systems. Particularly, what is the financial viability of such an investment, considering that most States are already experiencing serious difficulties in implementing the currently required terrestrial-based facilities and services?

Yet, the widespread use of GPS for navigational purposes world-wide is an undisputed reality, not to say an undeterred monopoly, and market dominance is the word of the day in the U.S. government. In this context, its early implementation is being pushed forward, and airlines do expect to see returns for the investments in airborne equipment they have already made. A reminder to the inattentive: no liabilities will arise to any State under the Convention for the unauthorised use of GPS signals in its sovereign airspace.

In this context, the lack of any legal instrument addressing the liability of the signal-in-space provider has been declared by many States to be an insurmountable obstacle to the implementation of the systems. In the absence of an appropriate recourse action mechanism, Article 28 States and other potential defendants are extremely concerned about resulting in having to compensate for damages which other parties may be partly or totally responsible for. Particularly uneasy about the application of the doctrine of sovereign immunity, they fear it might render court action against the U.S. and the Russian Federation, or any other State or entity providing GNSS signals, facilities and services, in countries other than their home States difficult or even impossible,[14] in the sense that they might "refuse" to appear before the court seized of the case in a foreign jurisdiction.

In brief, with the introduction of the GNSS, the legal complexities which may arise in the event of an accident are profoundly exacerbated by the multiplicity of actors involved. Even though a variety of compensation channels exists and may be considered reasonably adequate, the lack of uniformity in the numerous applicable individual legal regimes may result in serious conflicts of law and jurisdiction. Several layers of interconnected liabilities can be expected to further complicate and extend legal proceedings, and victims might need to engage in numerous parallel and consecutive legal actions with no guarantees as to the recovery of the full value of the damage.[15]

In view of all the above-mentioned, and taking into consideration that the process for the adoption and entry into force of an amendment to the Chicago Convention, which could clarify the matter of the extent of responsibility, may extend endlessly into time, the international community is left with only three viable alternatives, with legal and institutional implications, as follows:

Additional Legal Arrangements

In assuming that a signal-in-space provider finally accepts to formalise its relationship with user States, legal arrangements might be entered into whereby the adequate delegation of duties shall be made. From a private international law perspective, non-performance would constitute a breach of contract giving rise to liability. Thus, while providing the necessary guarantees as regards the availability, continuity, accuracy, reliability and integrity of the systems, it would make it possible to clearly identify the extent of responsibility for both foreign provider and Article 28 States, and therefore allow for the proper allocation of liabilities in case of damage.

Whereas the approach would allow for speedy implementation of the systems, the primary commercial aspect of GNSS services would make individual parties free to negotiate whatever terms and conditions they saw fit, thus contributing to the complete lack of uniformity, especially by reason of the great number of contracts which would need to be concluded world-wide. In this respect, a model contract adopted by the relevant ICAO bodies might be useful. Still, it could not serve as a substitute for the whole legal framework, since it would not address the long-term GNSS in its entirety.[16]

Hence, notwithstanding the odds, at present, against the successful outcome of the above alternative, consideration should continue to be given to the establishment of an appropriate global legal framework to govern the operation and availability of future GNSS, which should especially allow for full participation of all interested parties in the operation and control of the systems. Such a legal framework, however, should not be limited to GNSS only, but also be extended to other aspects of the CNS/ATM systems.[17]

Addressing liability through a contractual framework[18] initially established at a regional level might be particularly useful as an interim solution. Here again, transparency would help identify the extent of responsibility for the different actors in the provision, operation and use of the GNSS services at each stage of the "contractual chain", in accordance with individual performance criteria established therein. In case of an accident, channelling of liability would eventually trace it to the party whose actions or omissions had been the cause of the damage. The flexibility of these contractual arrangements would not only fit in with the evolution of technology, but also contribute to the development of the global long-term legal framework through the comparison of regional solutions.

Yet, recalling the additional legal complexities and procedural problems which may arise in the event of a GNSS related accident, an international convention[19] under the aegis of ICAO to regulate the matter in a simple, clear and straightforward manner definitely remains the best possibility envisaged for the long-term. Taking into consideration the recommendations of the LTEP, and incorporating or further developing the fundamental principles contained in the Charter and the Council Statement, such an instrument would allow for the direct allocation of liabilities between all actors involved, while ensuring prompt, adequate and effective compensation.

A Civil System

At present, there is no indication that any sort of binding international agreement will be concluded in the short-term concerning the responsibility for the provision of the primary signal-in-space. To be precise, it is widely-known that neither the U.S. nor the Russian Federation have any intention whatsoever of solely assuming the burden of world-wide responsibility for a service they provide free of direct user charges to the international community.

In spite of the fact that, in the event of a GNSS failure related accident, "the relevant rules of liability will apply and the signal providers will be held responsible through recourse to the laws of the relevant State",[20] scepticism prevails. Most States feel there is still some cause for concern and are not prepared, at their own risk, to implement a system, the core element of which is outside their sovereign control, solely relying on the good faith of the provider of the signal-in-space.

In this regard, an exclusively civil, internationally operated and controlled GNSS with the capability of delivering a global service that would meet all RNP requirements, remains the best alternative envisaged, as well as the ultimate goal in the evolutionary institutional path for the future GNSS. Its feasibility, however, will be dictated by the financial means and the political will of the international community, and yet be spurred by the urgency of a practical solution.[21] A regional system could certainly function as a starting-point.

The optimum design architecture of a future civil system will have to satisfy many user applications apart from civil aviation. Different levels of safety and performance will be required, and will have a direct impact on each user's share of the cost of developing and operating the multi-modal system. Accordingly, the system must be need-driven to be commercially attractive and financially justifiable. Whether the international civil aviation community, as one of the most demanding users, will be disposed to bear the financial implications of having such a system providing sole-means navigation for all operations is yet to be seen. For all purposes, representing only a minor share of satellite navigation users, civil aviation users should not pay for more than their fair share of the costs of GNSS provision.[22]

In particular, a civil system should evolve from the existing elements, maintaining full interoperability therewith in order to enable a planned and cost-effective transition, which would allow for the gradual amortisation of the investment made, while ensuring the protection of investment in the present air navigation systems, not rendering available technology and useful equipment immediately obsolete.[23]

As for the alleged difficulties in generating revenue[24] from such a system whilst signals are already provided for civil use free of charge, it is here submitted that not far is the day when the guarantees and security offered by an internationally controlled civil system will prove sufficient drive against any military and monopolistic system with no other legal or institutional guarantees.

In this scenario, the world has followed with much interest the definition phase of Galileo,[25] the new generation European satellite system which is forecast to become operational in 2008. Opened to all interested partners, it is expected to play an important role in future GNSS.

A Provisional Solution

Whereas the development of an internationally controlled civil system decisively remains the ultimate institutional goal for future GNSS, and an international convention is the long-term solution for the GNSS legal framework which will instil the necessary confidence, practical considerations might provisionally dictate or, at least, reasonably persuade otherwise.

Prompt action is required so that the international civil aviation community can reap early benefits from the implementation of the CNS/ATM systems. Technologically feasible and economically viable, the systems will bring greater safety, improved accuracy and regularity, as well as increased capacity, economies and efficiency. Yet, where provider and user States appear to be at a total deadlock, legal and institutional concerns have brought implementation to a standstill.

In this regard, further work on the complex legal aspects should not delay the implementation of the systems.[26] Law typically follows technological progress. Experience in different areas indicates that future developments of technology and a clearer conception of the characteristics of the long-term GNSS might actually be the ones to present practical solutions to eventual legal problems, and thus contribute to a consensus in the development of an appropriate long-term legal framework.[27]

World-wide inactivity might also reflect upon the availability of the 1559 to 1610 MHz band, the core frequency for supporting present and future GNSS operations, and might serve as a strong argument against the exclusive allocation of the spectrum to the Aeronautical Radionavigation Service and the Radionavigation Satellite Service. However, sharing of GNSS frequency bands with other radiocommunication services is not feasible. A matter of great urgency, therefore, is the need to ensure their absolute continuous protection. International co-operation is essential in this regard, as is the political will of States to move forward with the implementation of the systems.[28]

Finally, it may be said that, at least at present, the very success of the early implementation of the CNS/ATM systems is largely dependent upon the degree of good faith with which promises made by the U.S. and the Russian Federation are kept so that confidence placed upon them might prevail in the relations between provider and user States. Nevertheless, it may constitute but a provisional solution, which will definitely not preclude any future or concomitant action as regards the above-mentioned ideal legal and institutional alternatives. In any respect, ICAO should retain its co-ordinating role in the planning, development and implementation of the systems.

The technology is ready and waiting. Procrastination might lead to progress stagnation and obsolescence. The challenge is to act decisively and in time.

The timing is now.

Notes

1   See especially, Chapter 7, Implications of Article 28 at 124ff.
2   See ICAO, *Air Navigation Services and Economics Panel, Report on Financial and Related Organisational and Managerial Aspects of Global Navigation Satellite System (GNSS) Provision and Operation*, ICAO Doc.9660 (May 1996) at para.2.6.1.
3   See ICAO, *Report of the Panel of Experts on the Establishment of a Legal Framework with regard to GNSS,* ICAO Doc.LTEP/1 (23 December 1996)   at para.3:15 [unpublished].
4   A.Kotaite, ICAO's Role with Respect to the Institutional Arrangements and Legal Framework of Global Navigation Satellite System (GNSS) Planning and Implementation (1996) XXI:II *Annals of Air and Space Law* 195 at 201 [hereinafter Kotaite].
5   K.O.Rattray, "Legal and Institutional Challenges for GNSS – The Need for Fundamental Obligatory Norms" (ICAO World-wide CNS/ATM Systems Implementation Conference, Rio de Janeiro, 14 May 1998) at 4.
6   See ICAO, *Report of the First Meeting of the Secretariat Study Group on Legal Aspects of CNS/ATM Systems*, ICAO SSG-CNS/I-Report (9 April 1999) at para.3.8.8.
7   For detailed information on GPS and GLONASS, see Chapter 2.
8   For a comprehensive review on the legal significance of the ICAO SARPs, see Chapter 5 at 80ff.
9   See especially, Chapter 5, The Exchange of Letters, at 83.
10   U.S., *Treaties and Other International Agreements: The Role of the U.S. Senate, A Study Prepared for the Committee on Foreign Relations* (U.S. Senate, 103d Cong., 1st Sess., Nov. 1993) at xvi.
11   *Ibid.* at xxxvii-xxxviii.
12   See Chapter 2 at 37.
13   For a comprehensive review on the possible use of GNSS as the sole means of navigation, unlawful interference and related concerns, see Chapter 2 at 50ff.
14   See ICAO, *Report of the Second Meeting of the Secretariat Study Group on the Legal Aspects of CNS/ATM Systems*, ICAO C-WP/11190 (22 November 1999) at para.2.1.3 [unpublished]. For a study on the liability of the U.S. government under the Federal Tort Claims Act, see Chapter 7 at 128.
15   See especially ICAO, *Second Meeting of the Secretariat Study Group on Legal Aspects of CNS/ATM Systems*, ICAO SSG-CNS/2 (20-21 October 1999), "GNSS Liability: An Assessment", ICAO Doc.SSG-CNS/I-WP/4 (4 October 1999), by F. Schubert, presented by R.D.van Dam at 16-19.
16   See Chapter 5, Checklist of Items at 82.
17   See ICAO, *World-wide CNS/ATM Systems Implementation Conference Report*, ICAO Doc.9719 (May 1998) at 5.1.10 [hereinafter *WW/IMP Report*].
18   For more information on the contractual framework approach, see Chapter 6 at 106 and Chapter 7 at 139.
19   See especially, Chapter 6, The User States' Perspective at 103.
20   Kotaite, *supra* note 4 at 203.
21   See Chapter 7, Future Operating Structures, notes 257-258 and accompanying text.
22   For more information on the issue of cost recovery, see Chapter 7 at 146.

23 See Chapter 5, note 41 and accompanying text. For a practical example concerning Galileo, see also N.Warinsko, "Ambitious Project Would Involve Europe in New Generation of Satellite Navigation Services" (1999) 54:9 *ICAO Journal* 4 at 5 [hereinafter Warinsko]. See also Chapter 2 at 45ff.

24 For a comprehensive review on the financing strategy set for Galileo, see Chapter 2 at 47-49.

25 For a detailed review on Galileo, see Chapter 2 at 45ff.

26 See *WW/IMP Report, supra* note 17, Recommendation 5/3.

27 See Chapter 6 at 103ff.

28 On the issue of the GNSS frequency allocation, see Chapter 3.

Bibliography

ICAO Documents

Council

ICAO, *Council, 11ᵗʰ Session, Proceedings of the Council - II, Principles Governing the Reporting of Differences from ICAO Standards, Practices and Procedures*, ICAO Doc.7188 – C/828 *(1950)*.

ICAO, *Council, 110ᵗʰ Session*, ICAO Doc.9527 – C/1078, C-Min 110 and C-Min 110/9 (1983).

ICAO, Council, 154ᵗʰ Session, "Study on a Proposal for an International Aeronautical Monetary Fund", Appendix, "The Funding Mechanism of the ICAO Technical Co-operation Programme and Some Examples of Multilateral Mechanisms", ICAO C-WP/10840 (30 April 1998).

ICAO, *Council, 154ᵗʰ Session,* "Study on a Proposal for an International Aeronautical Monetary Fund", ICAO C-WP/10880 (15 May 1998).

ICAO, *Council, 155ᵗʰ Session, 7ᵗʰ Meeting*, ICAO C-Min 155/7 (22 February 1999).

ICAO, *Council, 156ᵗʰ Session*, "Policy on the Future Use of the Global Positioning System", ICAO Doc.C-WP/11097 (9 March 1999).

ICAO, *Council, 156ᵗʰ Session, 11ᵗʰ Meeting*, ICAO C-DEC 156/11 (15 March 1999).

ICAO, *Council, 156ᵗʰ Session*, "Use of GNSS as a Sole Means of Navigation", ICAO C-WP/11051 (5 February 1999).

ICAO, *Council, 156ᵗʰ Session, 2309ᵗʰ Report to the Council by the President of the Air Navigation Commission*, ICAO Doc.C-WP/11057 (8 March 1999).

ICAO, *Council, 158ᵗʰ Session*, "Study on an International Aeronautical Fund", ICAO C-WP/11235 (24 September 1999).

ICAO, *Council, 161ˢᵗ Session*, "International Financial Facility for Aviation Safety", ICAO C-WP/11467 (20 October 2000).

ICAO, *Council, 161ˢᵗ Session, 12ᵗʰ Meeting*, ICAO C-DEC 161/12 (30 November 2000).

ICAO, *Council, 162ⁿᵈ Session*, "Possible Relationship between an International Financial Facility for Aviation Safety (IFFAS) and the Global Aviation Safety Plan (GASP), ICAO C-WP/11545 (6 February 2001).

ICAO, *Council, 163ʳᵈ Session*, "Draft Assembly Working Paper - Mobilisation of Funds for Civil Aviation", ICAO C-WP/11591 (8 May 2001).

ICAO, *Council, 163ʳᵈ Session*, "Draft Assembly Working Paper - Mobilization of Funds for Civil Aviation", Attachment to C-WP/11591 (8 May 2001).

ICAO, *Council, 163ʳᵈ Session*, "Draft Assembly Working Paper on the Establishment of a Mechanism for an International Financial Facility for Aviation Safety (IFFAS), Discussion Paper n. 1 related to C-WP/11651 (20 June 2001).

ICAO, *Council, 163ʳᵈ Session, 9ᵗʰ Meeting*, ICAO C-DEC 163/9 (14 June 2001).

ICAO, *Council, 163ʳᵈ Session, 19ᵗʰ Meeting*, ICAO C-DEC 163/19 (29 June 2001).

ICAO, *Annual Report of the Council –2000*, ICAO Doc.9770 (2000).

ICAO, *Council, Report of the Second Meeting of the ALLPIRG/Advisory Group*, PRES AK/594 (11 March 1998).

Assembly

ICAO, *Assembly, 31ˢᵗ Session*, "Implementation of ICAO Standards and Recommended Practices", ICAO Doc.A-31 WP/56 (1 August 1995).

ICAO, *Assembly, 31ˢᵗ Session, Executive Committee*, "Strategic Action Plan", ICAO A31-WP/73.

ICAO, *Assembly, 32ⁿᵈ Session*, CD-ROM (Montreal, 1998), *Establishment of an ICAO Universal Safety Oversight Audit Programme*, Res. A32-11.

ICAO, *Assembly, 32ⁿᵈ Session, Consolidated Statement of ICAO Continuing Policies and Associated Practices Related Specifically to Air Navigation*, Res. A32-14.

ICAO, *Assembly, 32nd Session*, CD-ROM (Montreal, 1998), *ICAO Global Aviation Safety Plan (GASP)*, Res. A32-15.

ICAO, *Assembly 32nd Session*, CD-ROM (Montreal, 1998), *Charter on the Rights and Obligations of States Relating to GNSS Services*, Res. A-32-19.

ICAO, *Assembly 32nd Session*, CD-ROM (Montreal, 1998), *Development and Elaboration of an Appropriate Long-term Legal Framework to Govern the Implementation of GNSS*, Res. A-32-20.

ICAO, *Report of the 32nd Session of the ICAO Assembly, Legal Commission*, ICAO Doc. A32/LE (September-October 1998).

ICAO, *Assembly, 32nd Session, Legal Commission, Recommendations of LTEP*, ICAO Doc.A-32-WP/24, Appendix B.

ICAO, *Assembly, 32nd Session, Legal Commission*, "Progress in the Work of the Panel of Legal and Technical Experts on the Establishment of a Legal Framework with Regard to GNSS (LTEP)", ICAO Doc.A-32-WP/24, LC/3 (18 June 1998).

ICAO, *Assembly, 32nd Session, Executive Committee*, "Report on Financial and Organisational Aspects of the Provision of Air Navigation Services", ICAO Doc. A-32-WP/49, EX/18 (3 July 1998).

ICAO, *Assembly, 32nd Session, Executive Committee*, "Shortcomings and Deficiencies in the Air Navigation Field", ICAO Doc.A-32-WP/96, EX-41, Appendix (13 August 1998).

ICAO, *Assembly, 32nd Session, Executive Committee*, "Transition to the ICAO Universal Safety Oversight Audit Programme", ICAO Doc.A-32-WP/61 (6 July 1998).

ICAO, *Assembly, 33rd Session, Technical Commission*, "Results of the ITU Radiocommunication Conference 2000 (WRC-2000) and Preparation for Future WRCs", ICAO A33-WP/40 (28 June 2001).

Air Navigation Commission

ICAO, *Air Navigation Commission, Report on the Results of the ITU World Radiocommunication Conference (2000)*, ICAO AN-WP/7546 (09 June 2000).

ICAO, *Air Navigation Commission*, "Approval of a Draft Assembly Working Paper - Progress Report on the ICAO Global Aviation Safety Plan", ICAO AN-WP/7626 (20 February 2001).

Legal Committee

ICAO, *Report of the 28ᵗʰ Session of the ICAO Legal Committee*, ICAO Doc.9588 – LC/188 (1992).

ICAO, *Legal Committee, 28ᵗʰ Session,* Report of the Rapporteur on "The Institutional and Legal Aspects of the Future Air Navigation Systems", by Werner Guldimann, ICAO Doc.LC/28-WP/3-1 (24 January 1992).

ICAO, *Legal Committee, 28ᵗʰ Session,* "General Information and Comments Resulting From FANS (II)/3", ICAO Doc.LC/28-WP/3-5 (7 May 1992).

ICAO, *Report of the 29ᵗʰ Session of the ICAO Legal Committee*, ICAO Doc.9630 – LC/189 (1994).

ICAO, *Legal Committee, 29ᵗʰ Session,* Report of the Rapporteur on the "Consideration, with regard to global navigation satellite systems (GNSS), of the establishment of a legal framework", by Kenneth Rattray, ICAO Doc.LC/29-WP/3-1 (3 March 1994).

Panel of Experts on the Establishment of a Legal Framework With Regard to GNSS

ICAO, *Report of the First Meeting of the Panel of Experts on the Establishment of a Legal Framework with regard to GNSS,* ICAO Doc.LTEP/1 (23 December 1996) [unpublished].

ICAO, *Report of the Second Meeting of the Panel of Legal and Technical Experts on the Establishment of a Legal Framework with regard to GNSS,* ICAO Doc.LTEP/2 (3 November 1997)[unpublished].

ICAO, *Report of the Third Meeting of the Panel of Legal and Technical Experts on the Establishment of a Legal Framework with regard to GNSS,* ICAO Doc.LTEP/3 (9 March 1998)[unpublished].

ICAO, *Report of the First Meeting of the Working Group on GNSS Framework Provisions (Working Group II) of the Panel of Legal and Technical Experts on the Establishment of a Legal Framework with Regard to GNSS (LTEP)* (25 April 1997), ICAO LTEP/2-WP/3 (15 September 1997) [unpublished].

ICAO, *Report of the Second Meeting of the Working Group on GNSS Framework Provisions (Working Group II) of the Panel of Legal and Technical Experts on the Establishment of a Legal Framework with Regard to GNSS (LTEP)* (5 September 1997), ICAO LTEP/2-WP/4 (15 September 1997) [unpublished].

ICAO, *Report of the Third Meeting of the Working Group on GNSS Framework Provisions (Working Group II) of the Panel of Legal and*

Technical Experts on the Establishment of a Legal Framework with Regard to GNSS (LTEP) (12 February 1998), Appendix 3 to LTEP/3 Report [unpublished].

ICAO, *Panel of Experts on the Establishment of a Legal Framework With Regard to GNSS*, LTEP/1 (25-30 November 1996), "Different Types and Forms of the Long-Term Legal Framework For GNSS", ICAO Doc.LTEP/1-WP/5 (20 September 1996).

ICAO, *Panel of Experts on the Establishment of a Legal Framework With Regard to GNSS*, LTEP/1 (25-30 November 1996), "Inmarsat Satellite Navigation Services Institutional and Contractual Aspects", ICAO Doc.LTEP/1-WP/11 (29 October 1996).

ICAO, *Panel of Experts on the Establishment of a Legal Framework With Regard to GNSS*, LTEP/1 (25-30 November 1996), "Outline of the Role and Functions of a Multi-Modal European GNSS Agency and its Place Within the Regulatory Chain", ICAO LTEP/1-WP/16 (25 October 1996).

ICAO, *Panel of Experts on the Establishment of a Legal Framework With Regard to GNSS*, LTEP/2 (6-10 October 1997), "Liability Aspects of GNSS", ICAO Doc.LTEP/2-WP/6 (1 October 1997).

ICAO, *Panel of Experts on the Establishment of a Legal Framework with regard to GNSS, Working Group on GNSS Principles (Working Group I)*, LTEP-WG/I (10-14 March 1997), "Introductory Note", ICAO Doc.LTEP-WG/I-WP/2 (20 February 1997).

ICAO, *Panel of Experts on the Establishment of a Legal Framework with regard to GNSS, Working Group on GNSS Framework Provisions (Working Group II)*, LTEP-WG/II (22-25 April 1997), "Legal Aspects of GNSS Certification", ICAO Doc. LTEP-WG/II-WP/2 (18 March 1997).

ICAO, *Panel of Experts on the Establishment of a Legal Framework with regard to GNSS, Working Group on GNSS Framework Provisions (Working Group II)*, LTEP-WG/II (22-25 April 1997), "Liability Aspects of GNSS", ICAO Doc. LTEP-WG/II-WP/3 (18 March 1997).

ICAO, *Panel of Experts on the Establishment of a Legal Framework with regard to GNSS, Working Group on GNSS Framework Provisions (Working Group II)*, LTEP-WG/II (22-25 April 1997), "Future Operating Structures", ICAO Doc. LTEP-WG/II-WP/5 (18 March 1997).

ICAO, *Panel of Experts on the Establishment of a Legal Framework with regard to GNSS, Working Group on GNSS Framework Provisions*

(Working Group II), LTEP-WG/II (22-25 April 1997), "Liability Aspects of GNSS", ICAO Doc.LTEP-WG/II-WP/7 (18 April 1997).

ICAO, *Panel of Experts on the Establishment of a Legal Framework with regard to GNSS, Working Group on GNSS Framework Provisions (Working Group II)*, LTEP-WG/II (22-25 April 1997), "Legal Aspects of GNSS Certification and Liability", LTEP-WG/II-WP/8 (18 April 1997).

ICAO, *Panel of Experts on the Establishment of a Legal Framework with regard to GNSS, Working Group on GNSS Framework Provisions (Working Group II)*, LTEP-WG/II (22-25 April 1997), "Analysis of Liability Provisions in Existing International Conventions, Treaties and Other Relevant Instruments and Their Applicability to GNSS", ICAO Doc.LTEP-WG/II-WP/9 (18 April 1997).

ICAO, *Panel of Experts on the Establishment of a Legal Framework with regard to GNSS, Working Group on GNSS Framework Provisions (Working Group II)*, LTEP-WG/II(2) (2-5 September 1997), "Report of the Results of the Informal Survey Conducted by Working Group II", ICAO Doc.LTEP-WG/II(2)-WP/2 (14 August 1997).

Secretariat Study Group on Legal Aspects of CNS/ATM Systems

ICAO, *Report of the First Meeting of the Secretariat Study Group on Legal Aspects of CNS/ATM Systems*, ICAO SSG-CNS/I-Report (9 April 1999) [unpublished].

ICAO, *Report of the Second Meeting of the Secretariat Study Group on the Legal Aspects of CNS/ATM Systems*, ICAO C-WP/11190 (22 November 1999) [unpublished].

ICAO, *Report of the Third Meeting of the Secretariat Study Group on the Legal Aspects of CNS/ATM Systems*, ICAO SSG-CNS/3-Report (23 June 2000) [unpublished].

ICAO, *First Meeting of the Secretariat Study Group on Legal Aspects of CNS/ATM Systems*, ICAO SSG-CNS/I-IP/1 (April 1999).

ICAO, *Second Meeting of the Secretariat Study Group on Legal Aspects of CNS/ATM Systems*, ICAO SSG-CNS/2 (20-21 October 1999), "From Article 28 of the Chicago Convention to the Contractual Chain Solution", ICAO SSG-CNS/2 Flimsy No.1 (21 October 1999).

ICAO, *Second Meeting of the Secretariat Study Group on Legal Aspects of CNS/ATM Systems*, ICAO SSG-CNS/2 (20-21 October 1999), "Legal Rules in the United Kingdom Applicable to ATC", ICAO SSG-CNS/2-WP/1 (20 October 1999).

ICAO, *Second Meeting of the Secretariat Study Group on Legal Aspects of CNS/ATM Systems*, ICAO SSG-CNS/2 (20-21 October 1999), "The Liability System of the French Air Traffic Control", ICAO SSG-CNS/2-WP/2 (20 October 1999).

ICAO, *Second Meeting of the Secretariat Study Group on Legal Aspects of CNS/ATM Systems*, ICAO SSG-CNS/2 (20-21 October 1999), "The Law in Canada", ICAO SSG-CNS/2-WP/3 (4 October 1999).

ICAO, *Second Meeting of the Secretariat Study Group on Legal Aspects of CNS/ATM Systems*, ICAO SSG-CNS/2 (20-21 October 1999), "GNSS Liability: An Assessment", ICAO Doc.SSG-CNS/I -WP/4 (4 October 1999).

ICAO, *Second Meeting of the Secretariat Study Group on Legal Aspects of CNS/ATM Systems*, ICAO SSG-CNS/2 (20-21 October 1999), "Legal Rules in Italy Applicable to ATC", ICAO SSG-CNS/2-WP/5 (15 October 1999).

ICAO, *Second Meeting of the Secretariat Study Group on Legal Aspects of CNS/ATM Systems*, ICAO SSG-CNS/2 (20-21 October 1999), "U.S. Rules for Claims Against Air Traffic Control for Damages or Injury Resulting from Failure of Navigation Aids", ICAO SSG-CNS/2-WP/6 (15 October 1999).

ICAO, *Second Meeting of the Secretariat Study Group on Legal Aspects of CNS/ATM Systems*, ICAO SSG-CNS/2 (20-21 October 1999), "An Overview of the Legal Rules in Australia Applicable to Claims Against ATC", ICAO SSG-CNS/2-WP/7 (20 October 1999).

ICAO, *Fourth Meeting of the Secretariat Study Group on Legal Aspects of CNS/ATM Systems*, (April 1999), "Interference with CNS/ATM Systems – Enforcement in the U.S.", ICAO SSG-CNS/4-WP/2 (8 December 2000).

ICAO, *Fourth Meeting of the Secretariat Study Group on Legal Aspects of CNS/ATM Systems*, (April 1999), "Unlawful Interference with CNS/ATM – Australian Practice and Law", ICAO SSG-CNS/4-WP/4 (14 December 2000).

Global Navigation Satellite System Panel

ICAO, *Report of the Third Meeting of the Global Navigation Satellite System Panel*, GNSSP/3 (12-23 April 1999) [unpublished].

ICAO, *Global Navigation Satellite System Panel, 3rd Meeting* (12-23 April 1999), "Use of GNSS as Sole Means of Navigation", ICAO Doc.GNSSP/3-WP/29 (9 April 1999).

World-wide CNS/ATM Systems Implementation Conference

ICAO, *World-wide CNS/ATM Systems Implementation Conference Report,* ICAO Doc.9719 (May 1998).

ICAO, *World-wide CNS/ATM Systems Implementation Conference* (Rio de Janeiro, 11-15 May 1998), "ICAO Global Strategy for Training and Human Factors", ICAO WW/IMP-WP/13 (11 May 1998).

ICAO, *World-wide CNS/ATM Systems Implementation Conference* (Rio de Janeiro, 11-15 May 1998), "Specific Organisational Aspects Pertaining to the ICAO CNS/ATM Systems", ICAO WW/IMP-WP/15 (4 February 1998).

ICAO, *World-wide CNS/ATM Systems Implementation Conference* (Rio de Janeiro, 11-15 May 1998), "Organisational Forms of Air Navigation Services at the National Level", ICAO WW/IMP-WP/16 (6 February 1998).

ICAO, *World-wide CNS/ATM Systems Implementation Conference* (Rio de Janeiro, 11-15 May 1998), "Impact of Civil Aviation on States' Economies", ICAO WW/IMP-WP/19 (20 March 1998).

ICAO, *World-wide CNS/ATM Systems Implementation Conference* (Rio de Janeiro, 11-15 May 1998), "International Cost Recovery Policy", ICAO WW/IMP-WP/23 (2 March 1998).

ICAO, *World-wide CNS/ATM Systems Implementation Conference* (Rio de Janeiro, 11-15 May 1998), "Sources of Funds and Financing Mechanisms", ICAO Doc.WW/IMP-WP/25 (2 March 1998).

ICAO, *World-wide CNS/ATM Systems Implementation Conference* (Rio de Janeiro, 11-15 May 1998), "Assistance Requirements of States for CNS/ATM Implementation", ICAO WW/IMP-WP/27 (11 May 1998).

ICAO, *World-wide CNS/ATM Systems Implementation Conference* (Rio de Janeiro, 11-15 May 1998), "Human Factors Issues in CNS/ATM", ICAO WW/IMP-WP/30 (11 May 1998).

ICAO, *World-wide CNS/ATM Systems Implementation Conference* (Rio de Janeiro, 11-15 May 1998), "GNSS System Status and Standardisation in Progress", ICAO WW/IMP-WP/36 (11 May 1998).

ICAO, *World-wide CNS/ATM Systems Implementation Conference* (Rio de Janeiro, 11-15 May 1998), "Results of GNSS Assessment For Application in Approach, Landing and Departure", ICAO WW/IMP-WP/37 (11 May 1998).

ICAO, *World-wide CNS/ATM Systems Implementation Conference* (Rio de Janeiro, 11-15 May 1998), "Surveillance Systems", WW/IMP-WP/40 (11 May 1998).

ICAO, *World-wide CNS/ATM Systems Implementation Conference* (Rio de Janeiro, 11-15 May 1998), "Airborne Collision and Avoidance Systems", ICAO WW/IMP-WP/41 (11 May 1998).

ICAO, *World-wide CNS/ATM Systems Implementation Conference* (Rio de Janeiro, 11-15 May 1998), "MTSAT: Japan's Contribution to the Implementation of the ICAO CNS/ATM Systems in the Asia/ Pacific Regions", ICAO WW/IMP-WP/45 (11 May 1998).

ICAO, *World-wide CNS/ATM Systems Implementation Conference* (Rio de Janeiro, 11-15 May 1998), "EGNOS Space Based Augmentation Service to GPS and GLONASS", ICAO WW/IMP-WP/67 (11 May 1998).

Directors General of Civil Aviation Conference on a Global Strategy for Safety Oversight

ICAO, *Directors General of Civil Aviation Conference on a Global Strategy for Safety Oversight,* Conclusions and Recommendations, ICAO DGCA/97-CR 1 to 8.

ICAO, *Directors General of Civil Aviation Conference on a Global Strategy for Safety Oversight,* "Safety Oversight Today", ICAO DGCA/97-WP-1 (1 October 1997).

ICAO, *Directors General of Civil Aviation Conference on a Global Strategy for Safety Oversight,* "Results from the ICAO Safety Oversight Program", ICAO DGCA/97-WP-2 (1 October 1997).

ICAO, *Directors General of Civil Aviation Conference on a Global Strategy for Safety Oversight,* "Dealing with Confidentiality Issues", ICAO DGCA/97-WP-4 (2 October 1997).

ICAO, *Directors General of Civil Aviation Conference on a Global Strategy for Safety Oversight,* "Expansion of the ICAO Safety Oversight Programme to Other Technical Fields", ICAO DGCA/97-WP-6 (3 October 1997).

ICAO, *Directors General of Civil Aviation Conference on a Global Strategy for Safety Oversight,* "Relationship of the U.S. Federal Aviation Administration's International Aviation Safety Assessment Program to ICAO's Safety Oversight Program", ICAO DGCA/97-IP/1 (3 November 1997).

ICAO, *Directors General of Civil Aviation Conference on a Global Strategy for Safety Oversight,* "Safety Oversight, an International Responsibility", ICAO DGCA/97- IP/5 (20 October 1997).

ICAO, *Directors General of Civil Aviation Conference on a Global Strategy for Safety Oversight,* "The ICAO Safety Oversight Programme, A

Quality Assurance Approach to Safety", ICAO DGCA/97-IP/6 (23 October 1997).

Communications, Navigation and Surveillance/Air Traffic Management (CNS/ATM) Systems Implementation Task Force

ICAO, *Report of the First Meeting of the Communications, Navigation and Surveillance/Air Traffic Management (CNS/ATM) Systems Implementation Task Force*, CASITAF/1 (24-26 May 1994).

ICAO, *Report of the Second Meeting of the Communications, Navigation and Surveillance/Air Traffic Management (CNS/ATM) Systems Implementation Task Force*, CASITAF/2 (20-22 September 1994).

Special Committee on Future Air Navigation Systems

ICAO, *Report of the Fourth Meeting of the Special Committee on Future Air Navigation Systems (FANS)*, ICAO Doc.9524 - FANS/4 (2-20 May 1988).

ICAO, *Report of the Fourth Meeting of the Committee for the Monitoring and Co-ordination of Development and Transition Planning for the Future Air Navigation System (FANS PHASE II)*, ICAO Doc.9623 - FANS (II)/4 (15 September – 1 October 1993).

Conference on the Economics of Airports and Air Navigation Services

ICAO, *Report of the Conference on the Economics of Airports and Air Navigation Services* [hereinafter *ANSConf Report*], ICAO ANSConf-WP/115 (19 – 28 June 2000).

ICAO, "Study on Privatisation in the Provision of Airports and Air Navigation Services", ICAO ANSConf-WP/6, Appendix A.

ICAO, *Conference on the Economics of Airports and Air Navigation Services, "Pre-funding of Projects Through Charges"*, ICAO ANSConf-WP/15 (16 February 2000).

ICAO, *Conference on the Economics of Airports and Air Navigation Services*, "Joint Ventures in the Billing and Collection of Air Navigation Services", ICAO ANSConf-WP/22 (7 April 2000).

ICAO, *Conference on the Economics of Airports and Air Navigation Services*, "Forward Financing", ICAO ANSConf-WP/31 (24 December 1999).

ICAO, *Conference on the Economics of Airports and Air Navigation Services*, "The Allocation of GNSS Costs", ICAO ANSConf 2000-WP/65 (12 May 2000).

ICAO, *Conference on the Economics of Airports and Air Navigation Services*, "Revenue Diversion", ICAO ANSConf-WP/77 (5 June 2000).

Other Documents

ICAO, *Air Navigation Services and Economics Panel, Report on Financial and Related Organisational and Managerial Aspects of Global Navigation Satellite System (GNSS) Provision and Operation*, ICAO Doc.9660 (May 1996).

ICAO, *Checklist of Items to be Considered in Contracts for GNSS Signal Provision With Signal Providers in the Context of Long-term GNSS*, in ICAO Doc.9630 – LC/189 (1994).

ICAO, *Draft Agreement Between the International Civil Aviation Organisation (ICAO) and GNSS Signal Provider Regarding the Provision of Signals For GNSS Services*, in ICAO Doc.9630-LC/189 (1984).

ICAO, *Economics of Satellite-Based Air Navigation Services – Guidelines for Cost/Benefit Analysis of Communications, Navigation, Surveillance/Air Traffic Management (CNS/ATM) Systems*, ICAO Circ.257-AT/106.

ICAO, *Guidelines for the Introduction and Operational Use of the Global Navigation Satellite System*, ICAO Circ.267.

ICAO, *Global Air Navigation Plan for CNS/ATM Systems Executive Summary*.

ICAO, *Global Air Navigation Plan for CNS/ATM Systems*, version 1 (Montreal: ICAO, 1998) vols. 1 and 2.

Letter from D. Hinson, FAA Administrator, to A. Kotaite, President of ICAO Council (14 October 1994); Letter from A. Kotaite to D. Hinson (27 October 1994), ICAO State Letter LE 4/4.9.1-94/89, attachment 1 (11 December 1994).

Letter from N.P. Tsakh, Minister of Transport of the Russian Federation, to A. Kotaite, President of ICAO Council (4 June 1996), Letter from A. Kotaite to N.P. Tsakh (29 June 1996), ICAO State Letter LE 4/49.1-96/80 (20 September 1996).

ICAO, *Manual of the Regulation of International Air Transport*, ICAO Doc.9626 (1996).

ICAO, *Memorandum on ICAO, The Story of the International Civil Aviation Organisation*, 15th ed. (Montreal: ICAO, 1994).

ICAO, *Ninth Meeting of the Caribbean and South American Regional Planning and Implementation Group*, GREPECAS/9 (7-12 August 2000), "Recent Developments in the Modernisation of the Global Positioning System and U.S. Satellite Navigation Program Status".

ICAO, *Report of the World-wide Air Transport Conference on International Air Transport Regulation: Present and Future*, ICAO Doc.9644 (1994).

ICAO, *Safety Oversight Assessment Handbook*, 4th ed., 1997.

ICAO, *Statement of ICAO Policy on CNS/ATM Systems Implementation and Operation*, ICAO Doc.LC/29 - WP/3-2 (28 March 1994).

ICAO, *ICAO's Policies on Charges for Airports and Air Navigation Services*, ICAO Doc.9082/6 (2001).

ICAO, *Sixth Meeting of Directors of Civil Aviation - ICAO South American Region*, RAAC/6-IP/4.

ICAO, *The World of Civil Aviation, 1999 - 2002*, ICAO Circ.279– AT/116.

European Union Documents

EU, *Communication COM (2000) 750 final of 22 November 2000, Galileo*, [2000], http//www.galileo-pgm.org/indexrd.htm (date accessed: 5 January 2001).

EU, *Commission Communication of 10 February 1999, Galileo, Involving Europe in a New Generation of Satellite Navigation Services, Final Text*, G:\07\02\08\01-EN\final\text.doc [1999], http:/www.fma.fi/radionavigation/doc/galileo2.pdf (date accessed: 5 December 1999).

EU, *Commission Communication to the European Parliament, the Council and the Economic and Social Committee "Galileo – Involving Europe in a New Generation of Satellite Navigation Services*, Bulletin EU 1/2 – 1999 – Transport (5/23) (Brussels: EC, 25/05/1999).

EU, *Commission Working Document, Sec (1999) 789 final of 7 June 1999, Towards a Coherent European Approach for Space*, [1999], http://europa.eu.int/comm/jrc/space/com_doc_en.html (date accessed: 5 December 1999).

EU, *Communication COM (1998) 29 final of 21 January 1998, Towards a Trans-European Positioning and Navigation Network, Including a European Strategy for Global Navigation Satellite Systems (GNSS)* [1998] Bulletin EU ½ 1998, Transport (1/26).

EU, *Communication COM (1999) 54 final of 10 February 1999, Galileo, Involving Europe in a New Generation of Satellite Navigation Services* [1999] Bulletin EU 1/2 1999, Transport (5/23) at 1.3.169.

EU, *Council Resolution of 17 June 1999 on the Commission Communication on "Galileo, Involving Europe in a New Generation of Satellite Navigation Services"*, [1999] Bulletin EU 6-1999, Transport (2/9).

EU, *Council Resolution of 19 July 1999 on the Involvement of Europe in a New Generation of Satellite Navigation Services – Galileo – Definition Phase,* [1999] O.J.C. 1999/C 221/01.

EU, *Directive 85/374/EEC of 25 July 1985 on the Approximation of the Laws, Regulations and Administrative Provisions of the Member States Concerning Liability for Defective Products*, [1985] O.J.L. 210/29.

EU, *Galileo Working Document, Version 1.1 of 7 June 2000, Galileo Definition Phase – Initial Results* [2000], http://www.galileo-pgm.org/indexrd.htm (date accessed: 14 August 2000).

Other Government Documents

Canada, *State Immunity Act*, S.C. 1980-81-82-83, c. 95, s. 5, 6.

RAND Critical Technology Institute, *Global Positioning System. Assessing National Policies* (Santa Monica: Rand, 1995).

Russian Federation, *Directive of the President of the Russian Federation No. 38-rp* (18 February 1999), http://mx.iki.rssi.ru/SFCSIC/english.html (date accessed: 10 August 2000).

Russian Federation, Ministry of Defence, *GLONASS Interface Control Document,* version 4.0 (Moscow: Scientific Co-ordination Information Centre, 1998).

Russian Federation, Ministry of Defence, *Global Navigation Satellite System - GLONASS* (Moscow: Scientific Co-ordination Information Centre, 2000), http://mx.iki.rssi.ru/SFCSIC/english.html (date accessed: 12 August 2000).

Spain, *CNS/ATM Cost-Benefit Analysis for Spain: Final Report* (Aeropuertos Españoles y Navegación Aérea, 1996) vol. 1, 2.

UK, Department of the Environment, Transport and the Regions, *Consultation on the European Commission's Communication on Galileo, Involving Europe in a New Generation of Satellite Navigation Services COM (1999) 54 final* (April 1999), http://www.aviation.detr.uk.consult/galileo/index/htm (date accessed: 09 August 1999).

U.S., *Civilian Benefits of Discontinuing Selective Availability – Fact Sheet* (Department of Commerce, 1 May 2000).

U.S., *National Civilian GPS Services* (Washington, D.C., Department of Transportation, 21 March 2000).

U.S., *Global Positioning System Data and Information Files* (U.S. Naval Observatory, Automated Data Service), http://tycho.usno.navy.mil/gps.html (date accessed 12 August 2000).

U.S., *Statement by the President Regarding the United States' Decision to Stop Degrading Global Positioning System Accuracy* (White House, 1 May 2000).

U.S., *The Global Positioning System. Charting the Future* (Washington, D.C., National Academy of Public Administration and National Research Council, 1995) (Chair: J.R. Schlesinger).

U.S., *U.S. Global Positioning System Policy* (The White House Office of Science and Technology Policy and the National Security Council, 29 March 1996).

U.S., *The Federal Tort Claims Act*, 28 U.S.C. (1988).

U.S., *The Foreign Claims Act*, 10 U.S.C.A.(1996).

U.S., *The Military Claims Act*, 10 U.S.C.A.(1996).

U.S., *Global Positioning System Standard Positioning Service – Signal Specification*, 2nd ed. (The U.S. Coast Guard, 1995).

U.S., *Treaties and Other International Agreements: The Role of the U.S. Senate, A Study Prepared for the Committee on Foreign Relations* (U.S. Senate, 103d Cong., 1st Sess., Nov. 1993).

International Agreements and Conventions

Agreement on the Joint Financing of Certain Air Navigation Services in Greenland and the Faroe Islands, 1956, ICAO Doc.7726-JS/563.

Agreement on the Joint Financing of Certain Air Navigation Services in Iceland, 1956, ICAO Doc.7727-JS/564.

Charter of the United Nations and Statute of the International Court of Justice, 26 June 1945, 16 U.S.T.1134 (entered into force 24 October 1945).

Constitution of the International Telecommunications Union, Geneva, 22 December 1992 (entered into force 1 July 1994).

Convention for the Suppression of Unlawful Acts Against the Safety of Civil Aviation, 23 September 1971, ICAO Doc.8966 (entered into force 26 January 1973).

Convention for the Unification of Certain Rules Relating to International Carriage by Air, 12 October 1929, Schedule to the United Kingdom Carriage by Air Act, 1932; 22 and 23 Geo.5, ch.36 (entered into force 13 February 1933).

Convention for the Unification of Certain Rules for International Carriage by Air, 28 May 1999, DCW Doc.No.57 (not yet in force).

Convention on Damage Caused by Foreign Aircraft to Third Parties on the Surface, 7 October 1952, ICAO Doc.7364 (entered into force 4 February 1958).

Convention on International Civil Aviation, 7 December 1944, ICAO Doc.7300/6; UN Doc.15 U.N.T.S.295 (entered into force 4 April 1947).

Convention on International Civil Aviation, Annex 10, Aeronautical Telecommunications, vol. I-V.

Convention on International Civil Aviation, Annex 11, Air Traffic Services.

Convention on Offences and Certain Other Acts Committed on Board Aircraft, 14 September 1963, ICAO Doc.8364 (entered into force 4 December 1969).

Convention on the International Maritime Satellite Organisation (INMARSAT) 3 September 1976, 1143 U.N.T.S. 105 (entered into force 16 July 1979), amended 1985.

Convention on the International Liability for Damage Caused by a Space Object, 29 March 1972, 961 U.N.T.S. 187 (entered into force 1 September 1972)".

Declaration of Legal Principles Governing the Activities of States in the Exploration and Use of Outer Space", adopted unanimously on 13 December 1963.

ITU, *Radio Regulations* (1990), No. 17.

ITU, ITU WRC-2000 Provisional Final Acts, 2nd Edition (9 August 2000).

International Convention on the Establishment of an International Compensation Fund for Oil Pollution Damage, 18 December 1971(entered into force 16 October 1978), as amended 19 November 1976 and 25 May 1984 (not yet into force).

International Convention on Civil Liability for Oil Pollution Damage, 29 November 1969 (entered into force 19 June 1975) as amended 19 November 1976 (entered into force 8 April 1981) and 25 May 1984 (not yet into force).

Operating Agreement on the International Maritime Satellite Organisation (INMARSAT), 3 September 1976, 31:1 U.S.T.135 (entered into force 16 July 1979).

Treaty on the Principles Governing the Activities of States in the Exploration and Use of Outer Space, Including the Moon and Other Celestial Bodies, 27 January 1967, 610 U.N.T.S. 205 (entered into force 10 October 1967).

Vienna Convention on the Law of the Treaties, 23 May 1969, 1155 U.N.T.S. 331, Section 2 (entered into force 27 January 1980).

Books

Brownlie, I., *Principles of Public International Law* (Oxford: Clarendon Press, 1998).

Buergenthal, T., *Law-Making in the International Civil Aviation* Organisation (Syracuse, New York: Syracuse University Press, 1969).

Cheng, B., *General Principles of Law as Applied by the International Courts and Tribunals* (Cambridge: Grotius Publications, 1987).

Cheng, B., *The Law of International Air Transport* (London: Stevens, 1962).

Christol, C.Q., *The Modern International Law of Outer Space* (New York: Pergamon Press, 1982).

Diederiks-Verschoor, I.H.Ph., *An introduction to Air Law*, 5th rev. ed. (Deventer: Kluewer Law and Taxation, 1993).

Galotti Jr., V.P., *The Future Air Navigation System (FANS)* (Aldershot: Ashgate, 1997).

Gorove, S., *Developments in Space Law, Issues and Policies* (Dordrecht: Martinus Nijhoff, 1991).

Groenewege, A., *Compendium of International Civil Aviation*, 2nd ed. (Montreal: IADC, 1998).

Guldimann, W. and Kaiser, S., *Future Air Navigation Systems: Legal and Institutional Aspects* (Dordrecht: Martinus Nijhoff, 1993).

Henaku, B.D.K., *The Law on Global Air Navigation by Satellite: A Legal Analysis of the CNS/ATM System* (AST, 1998).

Hurwitz, B.A., *State Liability for Outer Space Activities in Accordance with the Convention on the International Liability for Damage Caused by a Space Object* (Dordrecht: Martinus Nijhoff,1992).

Lyall, F., *Law and Space Telecommunications* (Aldershot: Dartmouth, 1989).

Matte, N.M., *Aerospace Law: Telecommunications Satellite* (Toronto: Butterworths, 1982).

Matte, N.M., *Treatise on Air-Aeronautical Law* (Montreal: McGill University, 1981).

Reijinen, B.C.M., *The United Nations Space Treaties Analysed* (Gif-sur-Yvette Cedex, France: Frontières, 1992).

Rezek, J.F., *Direito Internacional Público* (São Paulo: Saraiva, 1996).

Schermers, H.G. and Blokker, N.M., *International Institutional Law: Unity Within Diversity*, 3rd ed. (The Hague: Nijhoff, 1995).

White, R.L. and White Jr., H.M., *The Law and Regulation of International Space Communication* (Boston: Artech House, 1988).

Articles

Abeyratne, R.I.R., "The Latin American Initiative Towards Funding the CNS/ATM Systems" (1998) 77:143 *The Aviation Quarterly* 151.

Abeyratne, R.I.R., "The Evolution from FANS to CNS/ATM and Products Liability of Technology Providers in the U.S." (1994) 43:2 *Zeitschrift sur Luftund Weltraumrecht* 156.

Al-Ghamdi, S.A., "Alternative Approach to Implementation of CNS/ATM Systems Would Impose User Charge" (1993) 48:3 *ICAO Journal* 19.

Allen, D.L., Haraldsdottir, A., Lawler, R.W., Pirotte, K., Schwab, R., "The Economic Evaluation of CNS/ATM Transition" (1999) 47:185 *Revue Navigation* 25.

Andrade, A.A.L., "Convenção de Montreal: Derradeira Esperança para o Transporte Aéreo Internacional" (1999) 78 *Revista Brasileira de Direito Aeroespacial* 2.

Baccelli, G.R., "La Responsabilità del Construttore Aerospaziale Secondo la Giurisprudenza Comparatistica e la Direttiva CEE in Materia di Responsabilità per Prodotto Difettoso (1990) XIV:2 *Diritto e Pratica dell'Aviazione Civile* 359.

Baccelli, G.R., "L'Unification Internationale du Droit Privé Aérien: Perspectives en Matière de Responsabilité des Transporteurs, des Exploitants des Aéroports et des Services de Contrôle de la Circulation Aérienne" (1983) VIII *Annals of Air and Space Law* 3.

Bartkowiski, M., "Responsibility for Air Navigation (ATM) in Europe" (1996) XXI:I *Annals of Air and Space Law* 45.

Bond, L., "Global Positioning Sense II: An Update" (1997) 39:4 J. ATC 51.

Bostwick, P.D., "Liability of Aerospace Manufactures: *MacPherson v. Buick* Sputters into the Space Age" (1994) 22 *Journal of Space Law* 75.

Carel, O., "La Protection des Usagers du GNSS Contre les Interruptions de Service" (1998) 46:182 *Revue Navigation* 213.

Carel, O. and Jonquière, J.L., "Les Spécifications des Systèmes Complexes et Leur Validation" (1999) 47:185 *Revue Navigation* 12.

Cheng, B., "International Responsibility and Liability for Launch Activities" (1995) XX:6 *Air and Space Law* 297.

Chiavarelli, E., "Satelliti e Sicurezza della Navigazione Aerea: Aspetti Giuridici e Ipotesi di Responsabilità" (1990) XIV:2 *Diritto e Pratica dell'Aviazione Civile* 383.

van Dam, R.D., "Recent Developments at the European Organisation for the Safety of Air Navigation (EUROCONTROL)" (1998) XXIII *Annals of Air and Space Law* 311.

Delrieu, A., "CNS/ATM: le Concept et le Système tel qu'Adoptés para L'OACI" (1995) 13 *Le Transpondeur* 4.

Diez D., and Nárdiz, M., "Un Estudio de Rentabilidad Sobre el CNS/ATM Realizado Por España Da Resultados Positivos a Nivel Nacional" (1998) 8:40 *Boletín Informativo Aital* 12.

Draghi, S., "Estimation du Coût de la Mise en Oeuvre du GNSS en Europe" (1996) 44:173 *Revue Navigation* 25.

Dupont, J., "Une Convention Internationale pour le GNSS" (1998) 36:1661 *Air and Cosmos Aviation International*.

Epstein, J.M., "Global Positioning System (GPS): Defining the Legal Issues of Its Expanding Civil Use" (1995) 61 *Journal of Air Law and Commerce* 243.

Ezor, J.I., "Costs Overhead: Tonga's Claiming of Sixteen Geostationary Orbital Sites and the Implications for U.S. Space Policy" (1993) 24 *Law and Policy in International Business* 915.

Finnsson,G., "Airports and Route Facilities: International Cost Recovery Policies and Their Applicability in the Framework of New Forms of Infrastructure Provision" (1994) XIX:II *Annals of Air and Space Law* 283.

FitzGerald, G.F., "ICAO and the Joint Financing of Certain Air Navigation Services" – Part I (1986) XI *Annals of Air and Space Law* 17.

FitzGerald, G.F., "ICAO and the Joint Financing of Certain Air Navigation Services" – Part II (1987) XII *Annals of Air and Space Law* 33.

Foster, W.F., "The Convention on International Liability for Damage Caused by Space Objects" (1972) *Canadian Yearbook of International Law* 137.

Fox, M.A., "ICAO Ready to Help Meet Global Training Needs Associated with the CNS/ATM Systems" (1995) 50:4 *ICAO Journal* 14.

Fukumoto, K. and Abe, K., "First of Several Japanese Satellites Designed for Aeronautical Use is Scheduled for Launch in 1999" (1998) 52:9 *ICAO Journal* 16.

Fukumoto, K. and Abe, K., "MTSAT: Japanese Contribution to the Implementation of ICAO CNS/ATM Systems in the Asia/Pacific Region" (1998) 46:184 *Revue Navigation* 442.

Hamalian, S.K., "Liability of the U.S. Government in Cases of Air Traffic Controller Negligence" (1996) XI *Annals of Air and Space Law* 58.

Hartl, P. and Wlaka, M., "The European Contribution to a Global Navigation Satellite System" (1996) 12:3 *Space Policy* 167.

Heijl, M.C.F., "Aviation Community Working on the Development of Infrastructure Needed to Support Free Flight" (1997) 52:3 *ICAO Journal* 7.

Heijl, M.C.F., "CNS/ATM Road Map for the Future" (1995) *IFALPA International Quarterly Review* 7 and (1994) 49:4 *ICAO Journal* 10.

Henaku, B.D.K., "The International Liability of the Space Segment Provider" (1996) XXIII:I *Annals of Air and Space Law* 145.

Henaku, B.D.K., "Legal Issues Affecting the Use of Navigation Systems" (1999) 47:187 *Revue Navigation* 312.

Huang, J., "ICAO Panel of Experts Examining the Many Legal Issues Pertaining to GNSS" (1997) 52:8 *ICAO Journal* 19.

Huang, J., "Sharing Benefits of the Global Navigation Satellite System Within the Framework of ICAO" (1996) 3:4 *International Institute of Space Law – Proceedings* 1.

Huang, J., "Development of the Long-Term Legal Framework for the Global Navigation Satellite System" (1997) XXII:I *Annals of Air and Space Law* 585.

Jakhu, R., "The Legal Status of the Geostationary Orbit" (1982) 7 *Annals of Air and Space Law* 333.

Jakhu, R.S., "International Regulation of Satellite Telecommunication" (1991), in *Legal Aspects of Space Commercialisation* (Tokyo: CSP Japan, 1992).

Jakhu, R.S., Remarks, "Developments in the International Law of Telecommunications: Strategic Issues for a Global Telecommunication Market" (1989) 83 *American Society of International Law – Proceedings* 385.

Jakhu, R.S., "The Evolution of the ITU's Regulatory Regime Governing Radiocommunication Services and the Geostationary Satellite Orbit" (1983) VIII *Annals of Air and Space Law* 381.

Jasentulyana, N., "The Role of Developing Countries in the Formulation of Space Law" (1995) XX-II *Annals of Air and Space Law* 95.

Johns, J.C., "Enhanced Capability of GPS and Its Augmentation Systems Meets Navigation Needs of the 21st Century" (1997) 52:9 *ICAO Journal* 7.

Johns, J.C., "Navigating the 21st Century with GPS" (1997) 39:3 *Journal of Air Traffic Control*. 34.

Kaiser, S., "A New Aspect of Future Air Navigation Systems: How Secondary Surveillance Radar Mode S Could Protect Civil Aviation" (1992) 41:2 *Zeitschrift sur Luftund Weltraumrecht* 154.

Kaiser, S., "Infrastructure, Airspace and Automation – Air Navigation Issues for the 21st Century" XX:1 (1995) *Annals of Air and Space Law* 447.

Kinal, G.V. and Ryan, F., "Satellite-based Augmentation Systems: The Need for International Standards" (1999) 52:1 *Journal of Navigation* 70.

Kotaite, A., "ICAO's Role with Respect to the Institutional Arrangements and Legal Framework of Global Navigation Satellite System (GNSS) Planning and Implementation" (1996) XXI:II *Annals of Air and Space Law* 195.

Kotaite, A., "Investment and Training Needs Among the Challenges Facing Developing Countries" (1993) 48:2 *ICAO Journal* 24.

Kries, W. V., "Some Comments on U.S. Global Positioning System Policy" (1996) 45:4 *Zeitschrift sur Luftund Weltraumrecht* 407.

Kuranov, V. and Iovenko, Y., "Capability and Performance Make GLONASS Suitable for Navigation in All Phases of Flight" (1997) 52:9 *ICAO Journal* 11.

Lagarrigue, I. and Bloch, J.D., "Le GNSS et Le Droit des États: l'Affrontement Entre États Fournisseurs et États Utilisateurs Lors de la Conférénce de Rio sur le CNS/ATM" (1998) 43:183 *Revue Navigation* 345.

Levy, S.A., "Institutional Perspectives on the Allocation of Space Orbital Resources: The ITU, Common User Satellite Systems and Beyond" (1984) 16 *Case Western Reserve Journal of International Law* 171.

Lim, C. and Elias, O., "The Role of Treaties in the Contemporary International Legal Order" (1997) 66 *Nordic Journal of International Law* 1.

Lyall, F., "Communications Regulation: The Role of the International Telecommunication Union" (1997) 3 *The Journal of Information, Law and Technology*. http://elj.warwick.ac.uk/jilt/commsreg/97_3lyal/lyall.TXT (date accessed: 3 December 1999).

Maniatis, D., "The Law Governing Liability for Damage Caused by Space Objects" (1997) XXII-I *Annals of Air and Space Law* 369.

Marchand, A.J., "Santos-Dumont: Pionnier de l'Aviation" (1996) 77:4 *AeroFrance* 4.

Matte, N.M., "The Chicago Convention, Where From and Where To, ICAO?" (1994) XXI:I *Annals of Air and Space Law* 371.

Mattews, S., "European Air Safety in the New Millennium", in World Market Series, *Business Briefing: European Civil Aviation and Airport Development* (World Markets Research Centre, 1999) 105.

McDonald, K. D., "Technology, Implementation and Policy Issues for the Modernisation of GPS and its Role in a GNSS" (1998) 51:3 *Journal of Navigation* 281.

Mein, D.T.E., "La Commercialisation du Système de Navigation Aérienne du Canada" (1998) 46:184 *Revue Navigation* 474.

Mendez, J.A., "Cuestiones Técnicas y Jurídicas sobre los Nuevos Sistemas de Comunicaciones en la Navegación Aérea" in *La Aviación Civil Internacional y el Derecho Aeronáutico Hacia el Siglo XXI* (Buenos Aires: ALADA, 1994) 161.

Michael, G.E., "Legal Issues Including Liability Associated With the Acquisition, Use and Failure of GPS/GNSS" (1999) 54:2 *Journal of Navigation* 246.

Milde, M., "Legal Aspects of Future Air Navigation Systems" (1987) XII *Annals of Air and Space Law* 87.

Milde, M., "Solutions in Search of a Problem? Legal Aspects of the GNSS" (1997) XXII:II *Annals of Air and Space Law* 195.

Milde, M., "The Chicago Convention – Are Major Amendments Necessary or Desirable 50 Years Later" (1994) XXI:I *Annals of Air and Space Law* 401.

Moffatt, J.F., "The Airport of the Future", in IATA, *Reinventing the Air Transport Industry - A Vision of the Future, Report of the Eight IATA High-Level Aviation Symposium* (1995) 102.

Moores, D., "RNP Implementation Demands Commitment and Careful Consideration of Many Issues" (1998) 53:2 *ICAO Journal* 7.

Mortimer, L., "1944 – 1994, A Half Century of Technological Change and Progress" (1994) 49:7 *ICAO Journal* 33.

Pace, S., "The Global Positioning System: Policy Issues for an Information Technology" (1996) 12:4 *Space Policy* 265.

Paylor, A., "Free Flight – The Ultimate Goal of CNS/ATM?" in ISC/ICAO, Integrating Global Air Traffic Management (London: ISC, 1997) 120.

Plave, M.E.F., "U.S. v. Varig Airlines: The Supreme Court Narrows the Scope of Government Liability under the Federal Tort Claims Act". (1985) 51 *Journal of Air Law and Commerce* 198.

Rattray, K.O., "The Changing Regulatory Environment, What Kind of World Will the Airlines be Flying In?" IATA, *Reinventing the Air Transport Industry - A Vision of the Future, Report of the Eight IATA High-Level Aviation Symposium* (1995) 22.

Rothblatt, M.A., "Satellite Communications and Spectrum Allocation" (1982) 76 *American Journal of International Law* 56 (LEXIS/NEXIS).

Sagar, D., "Recent Developments at the International Mobile Satellite Organisation (INMARSAT)" (1998) XXIII *Annals of Air and Space Law* 343.

Salin, P. A., "An Update on GNSS Before the Next ICAO Experts Meeting on the Legal and Technical Aspects of the Future Satellite Air Navigation Systems" (1997) XXII:I *Annals of Air and Space Law* 505.

Salin, P. A., "Regulatory Aspects of Future Satellite Air Navigation Systems (FANS) on ICAO's 50th Birthday" (1995) 44:2 *Zeitschrift sur Luftund Weltraumrecht* 172.

Schubert, F.P., "Organisations Régionales et Gestion de la Circulation Aérienne: Réflexion Critique sur le Régionalisme Européen" (1995) XX:I *Annals of Air and Space Law* 377.

Schubert, F. P., "Pilots, Controllers, and the Protection of Third Parties on the Surface" (1998) XXIII *Annals of Air and Space Law* 185.

Schubert, F.P., "Réflexions sur la Responsabilité dans le Cadre du GNSS" (1997) 45:180 *Revue Navigation* 417.

Spiller, J. and Tapsell, T., "Planning of Future Satellite Navigation Systems" (1999) 52:1 *Journal of Navigation* 47.

Spradling, K.K., "The International Liability Ramifications of the U.S. NAVSTAR Global Positioning System" (1990) 33 *Colloquium on the Law of Outer Space* 93.

Thompson, J.C., Comment, "Space for Rent, The International Telecommunications Union, Space Law and Orbit/Spectrum Leasing" (1996) 62 *Journal of Air Law and Commerce* 279.

Thomson-CSF, "Egnos: The Future European Navigation System" (1997) 41 *Prospace* 6.

Trempat, Y., "Les Projets GNSS: La Contribuition Européenne" (1996) 44:173 *Revue Navigation* 41.

Turner, L., "Transitioning to CNS/ATM – Tools to the Future" (1997) 39:3 *Journal of Air Traffic Control* 13.

Vidler, N., "Human Factors Aspects in CNS/ATM Systems" (1996) 38:3 *Journal of Air Traffic Control* 72.

Warinsko, N., "Du GPS au GNSS, Le Point sur la Situation Internationale" (1995) 13 *Le Transpondeur* 19.

Warinsko, N., "Ambitious Project Would Involve Europe in New Generation of Satellite Navigation Services" (1999) 54:9 *ICAO Journal* 4.

Weber, L. and Jakob, A., "Activities of the International Civil Aviation Organisation" (1996) XXI:II *Annals of Air and Space Law* 403.

Welch Pogue, L., "The International Civil Aviation Conference (1944) and Its Sequel: The Anglo-American Bermuda Air Transport Agreement (1946) – Appendix 1, "The Manifest Destiny of International Air Transport" (1994) XIX:I *Annals of Air and Space Law* 3.

Wilson, J., "The International Telecommunication Union and the Geostationary Orbit – An Overview" (1998) XXIII *Annals of Air and Space Law* 241.

Young, W.T., "Potential Interference on the Radio Spectrum Allocated for GNSS Needs Urgent Attention" (1996) 51:7 *ICAO Journal* 25.

Other Documents

Bond, L., "The GNSS Safety and Sovereignty Convention of 2000AD" (Global Airspace 99, Washington DC, 3 February 1999) [unpublished].

Costa Pereira, R.C., Address (42nd Air Traffic Control Association Annual Conference and Exhibits, 30 September 1997) 39:4 *Journal of Air Traffic Control* 56.

Costa Pereira, R.C., "Funding and Implementing Regional and Sub-regional Solutions in Africa" (African Aviation Conference and Exhibition 1999, Washington, 28 June 1999) [unpublished].

Donato, A.M., Address (Panel on the Establishment of an International Aeronautical Monetary Fund, Salvador, Brazil. 13 June 1994).

EU, Research and Development Sector, "Get Galileo to Set Pace in Satellite Navigation", (10 February 1999), http://www.eubusiness.com/rd/index.htm (date accessed: 5 December 1999).

European Tripartite Group, "Europe Pursuing a Broad Multimodal Satellite Navigation Programme as its Contribution to GNSS" (1997) 52:9 *ICAO Journal* 13.

Folchi, M., Address (Panel on the Establishment of an International Aeronautical Monetary Fund, Salvador, Brazil. 13 June 1994)". Global Satellite Navigation: From GNSS-1 to GNSS-2" (1997) 41 *Prospace* 2.

Huang, J., Comments on "GNSS Future Legal Issues", the Discussion Paper presented by P. B. Larsen (UNISPACE III) [unpublished].

IATA, *Reinventing the Air Transport Industry - A Vision of the Future, Report of the Eight IATA High-Level Aviation Symposium* (1995).

ICAO Secretariat, "Annual Review of Civil Aviation – 1999" (2000) 55: 6 *ICAO Journal* 7.

ICAO Secretariat, "Increased ATC Automation May be Inevitable to Handle Increasing Traffic and Data" (1993) 48:5 *ICAO Journal* 16.

ICAO Secretariat, "Highlights of the 32nd Assembly" (1998) 53:9 *ICAO Journal* 5.

ICAO Secretariat, "Human Factors and Training: Crucial Issues in CNS/ATM Implementation", *Transition, ICAO CNS/ATM Newsletter* 98/05, (Autumn 1998) 1.

ICAO Secretariat, "Business Cases Essential to CNS/ATM Systems Planning", *Transition, ICAO CNS/ATM Newsletter* 98/05 (Autumn 1998) 2.

ICAO Secretariat, "ICAO Examines Establishment of and International Aeronautical Fund", *Transition, ICAO CNS/ATM Newsletter* 98/05 (Autumn 1998) 3.

ICAO Secretariat, "ICAO Launches Global Air Navigation Plan for CNS/ATM Systems", *Transition, ICAO CNS/ATM Newsletter* 97/3 (Autumn 1997).

ICAO Secretariat, "Rio Lays Institutional and Financial Groundwork", *Transition, ICAO CNS/ATM Newsletter* 98/5 (Autumn 1998).

ICAO Secretariat, "Charter or International Convention? Legal Experts Debate", *Transition, ICAO CNS/ATM Newsletter* 98/5 (Autumn 1998).

ICAO Secretariat, "Significant majority of States need Help", *Transition, ICAO CNS/ATM Newsletter* 98/5 (Autumn 1998).

ITU, "International Telecommunication Union" in *Space Law: Applications, Course Materials* (Montreal: McGill University, 1997).

Janis, R., "Charges vs. Taxes – Which is which and why?" (Conference on the Economics of Airports and Air Navigation Services, Montreal, 19 June 2000), ICAO ANSConf-SP/3.1.

Jennison, M.B., "A Legal Framework for CNS/ATM Systems" (ICAO World-wide CNS/ATM Systems Implementation Conference, Rio de Janeiro, 14 May 1998).

Jennison, M.B., "The International Law of Satellite-Based Navigation Aids" (American Bar Association Forum on Air and Space Law, Montreal, 3 August 2000).

Kelly, T., Address (Panel on the Establishment of an International Aeronautical Monetary Fund, Salvador, Brazil. 13 June 1994).

Kesharwani, T.R., "Privatisation in the Provision of Airport and Air Navigation Services" (ICAO Airport Privatisation Seminar, Forum for the NAM/CAR/SAM Regions, Guatemala City, 13 December 1999).

Kotaite, A., Opening Address (8th IATA High-Level Aviation Symposium, 24 April 1995), in IATA, *Reinventing the Air Transport Industry - A Vision of the Future, Report of the Eight IATA High-Level Aviation Symposium* (1995).

Kotaite, A., Address (International Telecommunication Union World Radio Communication Conference 2000, 8 May 2000).

Larsen, P. B., "Future GNSS Legal Issues" (Third United Nations Conference on the Peaceful Uses of Outer Space, UNISPACE III, 19-30 July 1999). [unpublished].

Mamlouk, M., "Galileo" (American Bar Association, Forum on Air and Space Law, Montreal, 3 August 2000).

Milde, M., "Aviation Safety Standards and Problems of Safety Audits" (Soochow University Seminar, Taipei, 28 June 1997) [unpublished].

Milde, M., "The International Flight Against Terrorism in the Air" (Tokyo Conference, 3 June 1993) [unpublished].

Ministério da Aeronáutica, *Alberto Santos Dumont, The Father of Aviation,* (Brazil: Editorial Antártica, 1996).

Nordeng, T.V., "International Legal Impact on National Implementation of Global Navigation Satellite Systems (GNSS)" (ICAO World-wide CNS/ATM Systems Implementation Conference, Rio de Janeiro, 14 May 1998).

Quiroz, A., "ICAO Safety Oversight Programme – An Overview" (Senior Civil Aviation Management Course, Lecture, International Aviation Management Training Institute, 8 June 1999).

Rattray, K.O.", Legal and Institutional Challenges for GNSS – The Need for Fundamental Obligatory Norms" (ICAO World-wide CNS/ATM Systems Implementation Conference, Rio de Janeiro, 14 May 1998).

Rattray, K.O., "Economic Regulation – What is the Role of Government in a Changing World?" (Conference on the Economics of Airports and Air Navigation Services, Montreal, 19 June 2000), ICAO ANSConf-SP/7.2.

Razafy, J., Address (Panel on the Establishment of an International Aeronautical Monetary Fund, Salvador, Brazil. 13 June 1994).

Cases

Berkovitz v. U.S., 486 U.S. 531 (1988).

Bowden v. Korean Air Lines, 814 F. Supp. 592 (E. D. Mich., 1993).

Carlos Butterfield Case (U.S. v. Denmark) [1890] 2 Int. Arb.

Dalehite v. U.S., 346 U.S. 15 (1953).

Eastern Airlines, Inc. v. Union Trust Co., 221 F.2d (D.C.Cir. 1955).

Hays v. U.S., 899 F.2d 438 (5th Cir.1990).

Indian Towing Co. v. U.S. 350 U.S. 61 (1955).

In re Korean Air Lines Disaster of Sept. 1, 1983, 807 F. Supp. 1073 (S.D.N.Y. 1992).

In re Paris Air Crash of March 31, 1974, 399 F. Supp. 732 (Cal. 1975).

Park v. Korean Air Lines, 24 Av. Cas. (CCH) 17,253 (S.D.N.Y. 1992).

Smith v. U.S., 507 U.S. 197 (1993).

The Nuclear Tests Case (Australia v. France) [1974] ICJ Reports, 253.

The Wimbledon Case, Dissenting Opinion by Anzilotti and Huber, [1923] PCIJ. Rep. Ser. A. No. 1.

U.S. v. S.A. Empresa de Viação Aérea Rio Grandense (Varig Airlines), 467 U.S. 797 (1984).

Index

civil aviation authority (CAA)
  regulatory power, 78
  supervisory function, 78
  see safety oversight
civil aviation regulations, 78
  compliance with ICAO SARPs, 78
  enforcement, 77, 78
  see safety oversight
civil (GNSS) system, 111, 175, 176
civil law, 127, 128
CNS/ATM systems
  administration, 7, 90, 91, 140, 141, 141
  benefits of, 4, 20, 21, 177
  capacity, 4
  concept, development of, 4-8
  consistency with Chicago Convention, 73, 102-104
  description of, 22-28
  global co-ordination, 4
  global compatibility, 4, 5
  implementation, 4, 5, 112, 140, 141, 148, 177
  interoperability of elements, 4, 143
  legal implications, 71, 73, 74
  no legal obstacles, 73, 106
  planning, 4, 5, 108, 114
  see air traffic management, communications, cost recovery, financing, Global Plan, Global Navigation Satellite System, human factors, navigation, surveillance, Statement of ICAO Policy on CNS/ATM Systems Implementation and Operation
Coarse Acquisition (C/A) code, 38, 39
  see Global Positioning System
combatant activity exception, 131
commercial air transport operations
  certification, 77
  supervision, 77
  inspection, right of access for, 78
  see safety oversight
commercial arrangements, 2
  alliances, mergers and take-overs, 2
  airline consolidation, 2
commercial loans, 148
Commission of the European Union, *see* European Commission
communications
  air-ground, 17
  groung-to-ground, 18

radionavigation aids, 17
safety related and non-safety-related, 22
satellite communication 18, 42 - 44
voice and data communication 17, 18
CNS/ATM systems, 23
  aeronautical administrative communications, 22
  aeronautical mobile satellite services (AMSS), 22
  aeronautical operational control communications (AOC), 22
  aeronautical passenger communications (APC), 22
  aeronautical telecommunications network (ATN), 22
  ATS communications, 22
  benefits of, 22
  radio frequencies, 65, 66
  secondary surveillance radar (SSR), 22
  see radio frequency
common law, 127, 128
Communication, Navigation, Surveillance/Air Traffic Management Systems, *see* CNS/ATM systems
compatibility of regional arrangements with global planning and implementation, 114
compensation of damages, 122, 123, 124, 127, 133, 140, 173
compensation channels, 132-136, 174
  see liability
Conference on the Economics of Airports and Air Navigation Services, 142, 147
contributory negligence, 127, 128, 129, 133, 135
co-operation and mutual assistance, 113
congestion
  airport, 2
  airspace, 2
  groundside, 2
continuity of (GNSS) services, 23, 82, 84, 88, 110, 111, 172
contractual arrangements, *see* contractual framework
contractual chain, 105, 106, 175
  see contractual framework
contractual framework
  channelling of liability, 105, 124, 131, 132, 139, 175
  contractual chain, 105, 106, 175
  see checklist of items, model contract